THE 21ST-CENTURY BRAIN

Explaining, Mending and Manipulating the Mind

THE 21ST-CENTURY BRAIN

Explaining, Mending and Manipulating the Mind

Steven Rose

JONATHAN CAPE
LONDON

Published by Jonathan Cape 2005

2 4 6 8 10 9 7 5 3 1

First published in Great Britain in 2005 by
Jonathan Cape
Random House, 20 Vauxhall Bridge Road,
London SW1V 2SA

Random House Australia (Pty) Limited
20 Alfred Street, Milsons Point, Sydney,
New South Wales 2061, Australia

Random House New Zealand Limited
18 Poland Road, Glenfield,
Auckland 10, New Zealand

Random House South Africa (Pty) Limited
Endulini, 5A Jubilee Road, Parktown 2193, South Africa

The Random House Group Limited Reg. No. 954009
www.randomhouse.co.uk

A CIP catalogue record for this book
is available from the British Library

ISBN 0-224-06254-9

Papers used by The Random House Group Limited are natural,
recyclable products made from wood grown in sustainable forests;
the manufacturing processes conform to the environmental
regulations of the country of origin

Typeset in Guardi by Palimpsest Book Production Ltd,
Polmont, Stirlingshire

Printed and bound in Great Britain by
Clays Ltd, St Ives plc, Bungay, Suffolk

Contents

Acknowledgements

My first and most enduring acknowledgement, as always, is to the sociologist Hilary Rose, companion, often co-author and as often friendly critic, for more than forty years now. I feel her sceptical eye always on me as I write. Her capacity to puncture the overblown claims of neuroscience and genetics has, I hope, diminished my excesses just as it has enriched my understanding. I do not expect she will agree with all that I have written here, but will also know that it could never have been written without her.

More than thirty years ago, early in my career as a neuroscientist, I tried to summarise my own and my discipline's understanding of the brain and its relationship to mental activity in a book called either hubristically or with paradoxical intent (take your choice) *The Conscious Brain*. The numbers of those claiming to be neuroscientists have increased, probably by some three orders of magnitude, in the intervening period, and my capacity to assimilate and interpret their current understandings has certainly not been able to keep pace. None the less, now towards the end of my career, I have attempted a similar task, illuminated (I hope) by richer philosophical and biological understanding and certainly modulated by greater humility towards what is not known or unknowable by my science but might be better approached by other methods and other ways of knowing. Meantime my concerns over the increasingly imperialising claims of some amongst my neuroscientist and geneticist colleagues have sharpened as neuroscience segues seamlessly into neurotechnology and what have become known as 'neuroethical' concerns have grown. The attempt to encompass all these themes within a single volume, hopefully accessible to a wide readership, has been encouraged by Hilary's and my agent Kay McCauley, and by Will Sulkin at Cape.

The book draws on the experience and knowledge of collaborative research, conferences and discussions over the decades with many colleagues, both within the Brain and Behaviour Research Group at the Open University and the wider scholarly community, and it is impossible to do justice to all these sources, some of which I may not even be consciously aware of. As far as possible, they will find their contributions acknowledged in the reference list. However, specific chapters of the book have been read and I hope improved by Kostya Anokhin, Annette Karmiloff-Smith, Buca Mileusnic, John Parnavelas and Anya Tiunova. I am grateful for references and input from Sarah Norgate, Jim McGaugh, Susan Sara, Chris Yeo, Charles Medawar and Janice Hill and her Overload group in Edinburgh, from the book's editor at Oxford University Press in New York, Fiona Stevens, and a meticulous and thoroughly enjoyable editing collaboration with Jörg Hensgen at Cape. Roger Walker made the original drawings from my incoherent scrawls. Of course, errors and inter-pretation are my responsibility alone.

Readers should note that my discussion of the origins of life in Chapter 2 is dealt with more fully in my book *Lifelines* (Penguin, 1997; 2nd ed. Vintage, forthcoming); figures 3.2, 3.5, 3.6 and 3.7 have been redrawn from that book, which discusses development in a rather different context. For permission to reproduce illustrations I would also like to thank akg-images (Fig. 8.1) and Professor David Smith (Fig. 7.2). Fig. 2.9 is re-interpreted from PS Churchland and TJ Sejnowski, *The Computational Brain* (MIT Press, Cambridge, Mass, 1992).

CHAPTER 1

The Promise – and the Threat

'BETTER BRAINS' SHOUTED THE FRONT COVER OF A SPECIAL EDITION OF *Scientific American* in 2003, and the titles of the articles inside formed a dream prospectus for the future: 'Ultimate self-improvement'; 'New hope for brain repair'; 'The quest for a smart pill'; 'Mind-reading machines'; 'Brain stimulators'; 'Genes of the psyche'; 'Taming stress'. These, it seems, are the promises offered by the new brain sciences, bidding strongly to overtake genetics as the Next Big Scientific Thing. The phrases trip lightly off the tongue, or shout to us from lurid book covers. There is to be a 'post-human future' in which 'tomorrow's people' will be what another author describes as 'neurochemical selves'. But just what is being sold here? How might these promissory notes be cashed? Is a golden 'neurocentric age' of human happiness 'beyond therapy' about to dawn? So many past scientific promises – from clean nuclear power to genetic engineering – have turned out to be so perilously close to snake oil that one is entitled to be just a little sceptical. And if these slogans do become practical technologies, what then? What becomes of our self-conception as humans with agency, with the freedom to shape our own lives? What new powers might accrue to the state, to the military, to the pharmaceutical industry, yet further to intervene in, to control our lives?[1]

I am a neuroscientist. That is, I study how the brain works. I do this because, like every other neuroscientist, I believe that learning 'how brains work' in terms of the properties of its molecules, cells and systems, will also help us understand something about how minds work. This, to me, is one of the most interesting and important questions that a scientist – or indeed any other searcher after truth – can ask. Yet what I and my fellow neuroscientists discover provides more than mere passive knowledge of the world. Increasingly, as the *Scientific American* headlines suggest, this knowledge offers the prospect of sophisticated technologies for predicting, changing and controlling minds. My purpose in this book is to explore just how far the neurosciences' increasing capacity to explain the brain brings in its train the power to mend, modulate and manipulate the mind.

Of course, for many neuroscientists, to ask how the brain works is equivalent to asking how the mind works, because they take almost for granted that the human mind is somehow embodied within the 1500 grams of densely packed cells and connections that constitute the brain. The opening sentences of a book are no place to prejudge that question, which has concerned not merely science but philosophy, religion and poetry for millennia, and I will come back to it for sure in due course. For now, let me continue unpacking what it means to be a neuroscientist. In particular, I am interested in one of the most intriguing, important and mysterious aspects of how minds work: how we humans learn and remember – or, to be more precise, what are the processes that occur in our brains that enable learning and memory to occur. To approach this problem, I use a variety of techniques: because non-human animal brains work very much in the same way as do our own, I am able to work with experimental animals to analyse the molecular and cellular processes that occur when they learn some new skill or task, but I also use one of the extraordinary new imaging techniques to provide a window into the human brain – including my own – when we are actively learning or recalling.

It is this range – from the properties of specific molecules in a small number of cells to the electrical and magnetic behaviour of hundreds of millions of cells; from observing individual cells down a microscope to observing the behaviour of animals confronted with new challenges – that constitutes the neurosciences. It is also what makes it a relatively new research discipline. Researchers have studied brain and behaviour from the beginnings of recorded science, but until recently it was left to chemists to analyse the molecules, physiologists to observe the properties of ensembles of cells, and psychologists to interpret the behaviour of living animals. The possibility – even the hope – of putting the entire jigsaw together only began to emerge towards the end of the last century.

In response, the US government designated the 1990s as The Decade of the Brain. Some four years later and rather reluctantly, the Europeans declared their own decade, which therefore is coming to its end as I write these words. Formal designations apart, the huge expansion of the neurosciences which has taken place over recent years has led many to suggest that the first ten years of this new century should be claimed as The Decade of the Mind. Capitalising on the scale and technological success of the Human Genome Project, understanding – even decoding – the complex interconnected web between the languages of brain and those of mind has come to be seen as science's final frontier. With its hundred billion nerve cells, with their hundred trillion interconnections, the human brain is the most complex phenomenon in the known universe – always, of course, excepting the interaction of some six billion such brains and their owners within the socio-technological culture of our planetary ecosystem!

The global scale of the research effort now put into the neurosciences, primarily in the US, but closely followed by Europe and Japan, has turned them from classical 'little sciences' into a major industry engaging large teams of researchers, involving billions of dollars from government – including its military wing   and the pharmaceutical industry. The consequence is that what were once disparate fields – anatomy, physiology, molecular biology, genetics and behaviour – are now all embraced within 'neurobiology'. However, its ambitions have reached still further, into the historically disputed terrain between biology, psychology and philosophy; hence the more all-embracing phrase: 'the neurosciences'. The plural is important. Although the thirty thousand or so researchers who convene each year at the vast American Society for Neuroscience meetings, held in rotation in the largest conference centres that the US can offer, all study the same object – the brain, its functions and dysfunctions – they still do so at many different levels and with many different paradigms, problematics and techniques.

Inputs into the neurosciences come from genetics – the identification of genes associated both with normal mental functions, such as learning and memory, and the dysfunctions that go with conditions such as depression, schizophrenia and Alzheimer's disease. From physics and engineering come the new windows into the brain offered by the imaging systems: PET, fMRI, MEG and others – acronyms which conceal powerful machines offering insights into the dynamic electrical flux through which the living brain conducts its millisecond by millisecond business. From the information sciences come claims to be able to model computational brain processes – even to mimic them in the artificial world of the computer.

Small wonder then that, almost drunk on the extraordinary power of these new technologies, neuroscientists have begun to lay claim to that final terra incognita, the nature of consciousness itself. Literally dozens of – mainly speculative – books with titles permutating the term 'consciousness' have appeared over the last decade; there is a *Journal of Consciousness Studies*, and Tucson, Arizona hosts regular 'consciousness conferences'. I remain sceptical. This book is definitely not about offering some dramatic new 'theory of consciousness', although that particular ghost in the machine is bound to recur through the text. Indeed, I will try to explain why I think that as neuroscientists we don't have anything very much useful to say about that particular Big C, and why therefore, as Wittgenstein said many years ago, we would do better to keep silent.

The very idea of a 'consciousness conference' implies that there is some agreement about how such an explanation of consciousness should be framed – or indeed what the word even means – but there is not. The rapid expansion of the neurosciences has produced an almost unimaginable wealth of data, facts, experimental findings, at every level from the submolecular to that of the brain as a whole. The problem, which concerns me greatly, is how to weld together this mass into a coherent brain theory. For the brain is full of paradoxes. It is simultaneously a fixed structure and a set of dynamic, partly coherent and partly independent processes. Properties – 'functions' – are simultaneously localised and delocalised, embedded in small clusters of cells or aspects of the working of the system as a whole. Of some of these clusters, and their molecular specialisms, we have partial understanding. Of how they relate to the larger neural scene, we are still often only at the hand-waving stage.

Naming ourselves neuroscientists doesn't of itself help us bring our partial insights together, to generate some Grand Unified Theory. Anatomists, imaging individual neurons at magnifications of half a million or more, and molecular biologists locating specific molecules within these cells see the brain as a complex wiring diagram in which experience is encoded in terms of altering specific pathways and interconnections. Electrophysiologists and brain imagers see what, at the beginning of the last century, in the early years of neurobiology, Charles Sherrington described as 'an enchanted loom' of dynamic ever-changing electrical ripples. Neuroendocrinologists see brain functions as continuously being modified by currents of hormones, from steroids to adrenaline – the neuromodulators that flow gently past each individual neuron, tickling its receptors into paroxysms of activity. How can all these different perspectives be welded into one coherent whole, even before any attempt is made to relate the 'objectivity' of the neuroscience laboratory to the day-to-day lived experience of our subjective experience? Way beyond the end of the Decade of the Brain, and at halfway through the putative Decade of the Mind, we are still data-rich and theory-poor.

However, our knowledges, fragmented as they are, are still formidable. Knowledge, of course, as Francis Bacon pointed out at the birth of Western science, is power. Just as with the new genetics, so the neurosciences are not merely about acquiring knowledge of brain and mind processes but about being able to act upon them – neuroscience and neurotechnology are indissolubly linked. This is why developments occurring within the neurosciences cannot be seen as isolated from the socio-economic context in which they are being developed, and in which searches for genetic or pharmacological fixes to individual problems dominate.

It is clear that the weight of human suffering associated with damage or malfunction of mind and brain is enormous. In the ageing populations of Western industrial societies, Alzheimer's disease, a seemingly irreversible loss of brain cells and mental function, is an increasing burden. There are likely to be a million or so sufferers from Alzheimer's in the UK by 2020. There are certain forms of particular genes which are now known to be risk factors for the disease, along with a variety of environmental hazards; treatment is at best palliative. Huntington's disease is much rarer, and a consequence of a single gene abnormality; Parkinson's is more common, and now the focus of efforts to alleviate it by various forms of genetic therapy.

Whilst such diseases and disorders are associated with relatively unambiguous neurological and neurochemical signs, there is a much more diffuse and troubling area of concern. Consider the world-wide epidemic of depression, identified by the World Health Organisation (WHO) as *the* major health hazard of this century, in the moderation – though scarcely cure – of which vast tonnages of psychotropic drugs are manufactured and consumed each year. Prozac is the best known, but only one of several such agents designed to interact with the neurotransmitter serotonin. Questions of why this dramatic rise in the diagnosis of depression is occurring are rarely asked – perhaps for fear it should reveal a malaise not in the individual but in the social and psychic order. Instead, the emphasis is overwhelmingly on what is going on within a person's brain and body. Where drug treatments have hitherto been empirical, neurogeneticists are offering to identify specific genes which might precipitate the condition, and in combination with the pharmaceutical industry to design tailor-made ('rational') drugs to fit any specific individual – so called psychopharmacogenetics.

However, the claims of the neurotechnologies go far further. The reductionist fervour within which they are being created argues that a huge variety of social and personal ills are attributable to brain malfunctions, themselves a consequence of faulty genes. The authoritative US-based *Diagnostic and Statistical Manual* now includes as disease categories 'oppositional defiance disorder', 'disruptive behaviour disorder' and, most notoriously, a disease called 'Attention Deficit Hyperactivity Disorder' (ADHD) is supposed to affect up to 10 per cent of young children (mainly boys). The 'disorder' is characterised by poor school performance and inability to concentrate in class or to be controlled by parents, and is supposed to be a consequence of disorderly brain function associated with another neurotransmitter, dopamine. The prescribed treatment is an amphetamine-like drug called Ritalin. There is an increasing world-wide epidemic of Ritalin use. Untreated children are said to be likely to be more at risk of becoming criminals, and there is an expanding literature on 'the genetics of criminal and anti-social behaviour'. Is this an appropriate medical/psychiatric approach to an individual problem, or a cheap fix to avoid the necessity of questioning schools, parents and the broader social context of education?

The neurogenetic-industrial complex thus becomes ever more powerful. Undeterred by the way that molecular biologists, confronted with the outputs from the Human Genome Project, are beginning to row back from genetic determinist claims, psychometricians and behaviour geneticists, sometimes in combination and sometimes in competition with evolutionary psychologists, are claiming genetic roots to areas of human belief, intentions and actions long assumed to lie outside biological explanation. Not merely such long-runners as intelligence, addiction and aggression, but even political tendency, religiosity and likelihood of mid-life divorce are being removed from the province of social and/or personal psychological explanation into the province of biology. With such removal comes the offer to treat, to manipulate, to control. Back in the 1930s, Aldous Huxley's prescient *Brave New World* offered a universal panacea, a drug called Soma that removed all existential pain. Today's Brave New World will have a multitude of designer psychotropics, available either by consumer choice (so called 'smart' drugs to enhance cognition) or by state prescription (Ritalin for behaviour control).

These are the emerging neurotechnologies, crude at present but becoming steadily more refined. Their development and use within the social context of contemporary industrial society presents as powerful a set of medical, ethical, legal and social dilemmas as does that of the new genetics, and we need to begin to come to terms with them sooner rather than later. To take just a few practical examples: if smart drugs are developed ('brain steroids' as they have been called), what are the implications of people using them to pass competitive examinations? Should people genetically at risk from Alzheimer's disease be given life-time 'neuroprotective' drugs? If diagnosing children with ADHD really does also predict later criminal behaviour, should they be drugged with Ritalin or some related drug throughout their childhood? And if their criminal predisposition could be identified by brain imaging, should preventative steps be taken in advance of anyone actually committing a crime?

More fundamentally, what effect do the developing neurosciences and neurotechnologies have on our sense of individual responsibility, of personhood? How far will they affect legal and ethical systems and administration of justice? How will the rapid growth of human-brain/machine interfacing – a combination of neuroscience and informatics (cyborgery) – change how we live and think? These are not esoteric or science fiction questions; we aren't talking about some fantasy human cloning far into the future, but prospects and problems which will become increasingly sharply present for us and our children within the next ten to twenty years. Thus yet another hybrid word is finding its way into current discussions: neuroethics.

*   *   *

These, then, are some of the issues which I have been exploring over my forty-five years as a researching neuroscientist, and with which, finally, this book is trying to come to terms. Just what is the future of the brain? By which I mean, what hope of 'understanding' the brain do we now have? Can we assemble the four-dimensional, multi-level jigsaw of the brain in space and time that is required before we can even begin the real business of decoding the relationships of mind and brain? Or, better from my point of view, of learning the translation rules between these two very different languages? And what of the future of all our brains and minds in a world in which the manipulative neurotechnologies are becoming ever more powerful?

To approach these questions requires that I begin by attempting an almost impossible task of assessing the current state of the brain sciences: what we neuroscientists know – or think we know – about the wrinkled mass of tissue inside each of our own heads. I tried this once before, thirty years or so ago, when I wrote a book called *The Conscious Brain*. It seemed easy to me then, being younger and more naïve – but that was long before the knowledge explosion of the last decades, so things appeared simpler. An update would be impossible, even if it were desirable. What I am trying here is rather different. Let me explain why.

The great evolutionary biologist Theodosius Dobzhansky once said that nothing in biology makes sense except in the light of evolution. So a starting point in any attempt to understand today's human brains should be to locate them in evolutionary terms: how and why might brains have evolved? Never an easy question, as neither brains nor nervous systems, still less behaviour, can be detected in the fossil record. Nevertheless, we can draw some useful conclusions by studying existing organisms, both those with developed brains and those with seemingly less complex nervous systems and forms of behaviour. So this is where, with the next chapter, the book-proper starts.

Dobzhansky's aphorism, however, is only part of what we need in order to understand living organisms. As well as their evolutionary history, we need to understand their developmental history, the path from fertilised egg to adult organism with a developed brain and repertoire of behaviour. Organisms construct themselves, their brains and their behaviour, out of the raw material provided by their genes and the environmental context with which they interact – a view of the world sometimes called developmental systems theory, or autopoiesis – and so my account of brain development follows that of evolution to form Chapter 3. Then, as life cycles require endings as well as beginnings, in Chapter 7 I turn from early to later years, to the ageing brain and its discontents.

Those three chapters provide the basis on which I can turn to the crucial issues. What does it mean to be human? If our genes are 99 per cent identical to those of chimpanzees, if our brains are composed of identical molecules, arranged in pretty similar cellular patterns, how come we are so different? Becoming human, becoming a person, forms the themes of Chapters 4 and 5. And so, at last to the key question: what about 'the mind'? Where in the hundred billion nerve cells of the brain – if there at all – will we find anything approaching mind? Or is this a strictly meaningless question, imposed on us by the structure of our own history, as neuroscientists work within the Western tradition with its love of dichotomies, of mind and brain, nature and nurture, neurological and psychological . . . ?

Chapter 6 is my own best take on these questions. As philosophers enjoy pointing out, there are paradoxes about using our brains/minds to try to understand our brains/minds. Attempting to do so involves fretting about quite deep questions within the philosophy and sociology of science, about how we know what we know. And how we know what we know depends on an interplay between at least three factors. One is obvious: the material nature of the world itself. As a scientist I am inevitably a realist; the world exists independently of my attempts to interpret it, even though I can only know the world through my own sense organs and the models that these help construct inside my own head. But these models are indeed constructed; they are not simply perfect scaled-down images of the world outside my head – they are indeed shaped by my own evolution and development, and, inextricably, by the society and culture within which that development has taken place. The ways in which we conduct our observations and experiments on the world outside, the basis for what we regard as proof, the theoretical frameworks within which we embed these observations, experiments and proofs, have been shaped by the history of our subject, by the power and limits of available technology, and by the social forces that have formed and continue to form that history.

The reductionist philosophical traditions of Western science mould our approach to understanding, yet are often sorely tested by the complexity of the real world. As I have already said, nothing in that real world is more complex than our brains, and we must always try to be aware of the degree to which our ideas are both formed and restricted by our history. So, to stretch Dobzhansky still further, nothing in our understanding of living organisms and processes makes sense except in the light of evolution, development, and our social, cultural, technological and scientific history. My account, in Chapter 6, is illuminated, or obscured, by the light and shadow coming from these multiple sources.

Science is not simply about the passive contemplation of nature, but carries with it an interventive corollary. Despite doubts amongst earlier philosophers and physicians in many cultures, from the ancient Egyptians to the Chinese, about the nature of the brain and the location of the seat of the mind, for the last three hundred years explaining the brain is also about attempting to heal the sick mind. It is these attempts that provide the forerunners to the present and future neurotechnologies. To set the potential of these technologies into context, however, I begin, in Chapter 8, by considering the future of neuroscience itself – what we know, what we might know, and what I see as the inevitable limits to such knowledge – before turning, in Chapter 9, to the history of attempts to use neuroscientific knowledge to heal the mind. As for the technologies, in Chapter 10 I begin with two exemplary case studies, the hunt for cognitive enhancers – the so-called smart drugs – and the use of drugs to control schoolchildren's behaviour.

Finally, in Chapters 11 and 12 I consider the newest threats and promises of neurotechnology and their ethical challenges: eliminating unwanted behaviour or enhancing desired characteristics; reading and changing our minds; controlling dissent; predicting and modifying the future. Governments already speak of 'brain policy'. Manufacturers are becoming interested in 'neuroeconomics' and even 'neuromarketing'. Is such 'progress' inevitable? How can we as citizens help shape and direct the goals, methods and uses of neuroscience and neurotechnology over the coming decades? This is the democratic challenge with which my book ends: a challenge to the institutionalisation of that other increasingly fashionable hybrid word, 'neuroethics'. It is the end of the book, but not of the debate.

CHAPTER 2

The Past Is the Key to the Present

Once upon a time . . .

To BEGIN AT THE BEGINNING. IN THE BEGINNING THERE WAS . . . WHAT?
Genesis offers us a void, chaos, out of which God creates order. John
in the New Testament gospel has an alternative version – in the beginning was the Word. Perhaps present day cosmologists and physicists,
seeking for what they describe as a 'theory of everything' would be
content with such a Word, provided it took the form of a complex mathematical equation. We biologists, by contrast, have little time for physicists' theories of everything. Life is complex enough for us to handle.
None the less, many present-day molecular biologists, seeking an origin
for, if not the universe, then at least life, accept John's version. For them
the beginning is constituted by the four letters, A,C,G and T, that
comprise the alphabet of DNA (deoxyribonucleic acid). In the molecular biologist's gospel, these four letters – which stand for the molecules known as nucleotides: adenosine, cytosine, guanine and thymidine
– comprise life's alpha and omega. The beginning and end of all individual lives, and of life in general, lie in the perpetuation of particular
combinations of these famous letters. If I have to choose, however, I
would rather opt for the Old Testament version, in which Words are
relative latecomers on life's stage. First come primitive cells, even perhaps
organisms with something we might call behaviour – though not yet
with the nervous systems and brains to support such behaviour.

So, let's return to the Genesis version, the chaos of the slowly cooling planet Earth, four billion years or so ago, and propose a slightly less mystical account. No life then, and certainly no brains, so how might one have got from there to here, from then to now? Here's a scenario. I've assembled it from a variety of sources, some sure, some more speculative, but at any rate plausible enough to move us from an imagined world into one where evidence can begin to play a part.[1] Note that in my imagined world, unlike that of those who begin with an alphabet soup of ACG and T, chickens come before eggs – cells before genes. And even more relevantly to my purposes in this chapter, behaviour – that is, goal-directed actions by cells upon the external world – comes before brains and even before nervous systems.

To begin with, one needs chemistry. Life, however defined ('a dynamic equilibrium in a polyphasic system', as the biochemist Frederick Gowland Hopkins described it three-quarters of a century ago), engages the complex interactions and interconversions of chemicals built from carbon, hydrogen, oxygen and nitrogen. These small and large molecules float in a watery sea – we humans are 80 per cent water – salty with the ions of sodium, potassium, calcium, chlorine, sulphur and phosphorus along with a range of heavy metals. The first problem is to get from the inorganic chemistry of the cooling earth into the organic, carbon chemistry of sugars, amino acids and the nucleotides (building blocks of nucleic acids), and from there to the giant molecules: proteins, fats, DNA and RNA (ribonucleic acid), bringing phosphorus, sulphur and the rest in along the way. There are several ways in which, in the abiotic past, the simplest of these might have been synthesised. Which is the more likely depends on assumptions about the primitive Earth's atmosphere – in particular that, rather than being oxygen-rich as it is today, it resembled that of the other planets of the solar system, with little or no oxygen but much nitrogen, ammonia and carbon dioxide.[2] Chemically, this is a reducing rather than an oxidising environment, which means that it is more conducive to the synthesis of complex but relatively unstable molecules. Perhaps in this environment life could have begun in the oceans, sparked by violent electrical storms, the precursors of Mary Shelley's Frankenstein, plucking life from heaven; or perhaps, just as violently, in the hot cauldrons of volcanic eruptions, or the steaming vents in the deep oceans where strange single-celled forms of life, *Archaea*, have only recently been discovered, living, to paraphrase Shakespeare's *Richard II*, 'like vermin where no vermin else but only they have right to live'; or perhaps more tranquilly, in the drying, clayey ocean margins where land and water mingle – currently the most favoured hypothesis.[3]

Even more improbably, some suggest that life did not begin on Earth at all, but was 'seeded' from outer space. Amino acids, for instance, can be synthesised from water, methanol, ammonia and hydrogen cyanide, at temperatures only a few degrees above absolute zero – conditions known to pertain in interstellar space. Furthermore, although higher temperatures do speed up chemical reactions, the complex organic molecules once formed are much more stable at low than high temperatures. DNA, for instance, is stable for hundreds of thousands of years in cool high latitudes, but only for thousands in warm lower latitudes.

All these theories have their advocates, their experimental demonstrations of possibility. The idea of 'panspermia' – of seeding from outer space, not just of simple molecules but of DNA or even fully competent cells, championed by Francis Crick[4] – has especially captured some writers' imagination, though I find it the least convincing. In any event, such abiotic synthesis theories are not mutually exclusive and whether one or all of them may have been in play, the results add up to the consequence that over the first few hundreds of million years of Earth history the oceans and their margins become a thin minestrone of organic chemicals – amino acids, sugars, fatty acids, even nucleotides. However, to get from soup to life is still a huge step – easy enough to reverse in the kitchen, for sure, but we need to run this film forwards, not backwards.

Proto-cells

The clue, I believe, lies above all in one class of these organics – lipids, or oils. Pour oil on to the surface of water and it will either spread as a thin film or form a droplet. The droplet has an inside and an outside, a lipid membrane dividing the world in two:[5] a separation of something that might in due course become an organism from what will become the environment of that organism, the biologist's equivalent of the Genesis story in which God divides the waters from the land, creating structures. For such droplets have an interesting physico-chemical property: they are able to concentrate within themselves many of the organic chemicals in the environmental soup, along with ions like those of calcium (Ca^{2+}) and potassium (K^+). The droplet becomes a proto-cell, with an internal chemical constitution quite distinct from that outside it. This difference between internal and external, between what will in due course become self and non-self, is one of the primary character-istics of living organisms; and indeed fossilised cell-like structures can be found in rocks at least 3.5 billion years old.

Once these organics begin to concentrate, especially if fragments of clay and metal salts offering surfaces on which catalytic reactions can occur are also trapped within the proto-cells, then things can begin to speed up. More and more complex chemical interconversions can occur as a result of the newly available catalytic possibilities. Computer models can be built showing how such a set of mutually interacting chemicals will ultimately stabilise, forming a dynamic metabolic web. Within this web, larger and larger organic molecules begin to be formed. Some, strings of amino acids (proteins) or of nucleic acids (RNA), have catalytic properties themselves, and thus function as enzymes, even capable of catalysing their own synthesis, making it possible for them to accelerate or even direct further syntheses.[6]

It is here also that the ionic composition of the environment inside the proto-cell becomes important. It is not just that the ions are involved in many catalytic processes, but also that the internal concentration of the ions becomes very different from that outside the cell, being high in potassium and calcium and low in sodium. A simple lipid membrane won't achieve quite this specificity, because it cannot readily select what to permit to enter and what to exclude. In real cells the membrane is selective, semi-permeable, and quite fussy about which ions or mole-cules it allows to enter or leave. This is in part the result of not being a simple lipid but also containing a variety of proteins which 'float' in the lipid environment (so-called 'lipo-protein rafts'), and in some way these must have become incorporated into my lipid proto-cell membrane at a relatively early stage in life's history.

The consequence of this is profound. Push an electrode with a fine tip through the membrane of any modern living cell, from an Amoeba to a human blood cell or nerve cell (henceforward 'neuron'), place a second electrode on the external surface of the cell or the surrounding fluid, and connect the two through a voltmeter and you will record a

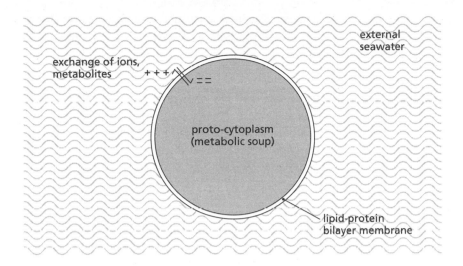

Fig. 2.1 *A proto-cell.*

potential difference (a voltage) between the inside and the outside of the cell of the order of 70–100 thousandths of a volt (millivolts), with the inside of the cell negative and the outside positive. The difference is caused by the uneven distribution of ions across the membrane, with the interior rich in negatively charged proteins, balanced by high levels of positively charged potassium (K^+) but with low sodium (Na^+) relative to the external medium. It may not sound like a lot, but it is worth remembering that the cell membrane is only one millionth of a centimetre across, so as a gradient the membrane potential is a hundred thousand volts per centimetre – quite a tidy charge! This membrane potential is at least as significant a defining feature of what makes life possible as any other aspect of cell structure and biochemistry, as will become apparent as my narrative develops.

To survive, to maintain their internal stability, grow and divide, my proto-cells need energy. To start with, of course, there is a reasonable amount of it about in the form of already synthesised molecules in the environmental soup. Energy can thus be obtained by capturing these molecules – sugars, for instance – sucking them into the proto-cells from the oceanic soup in which they float. Once inside they can be broken down – oxidised, the equivalent of burning – to release their trapped chemical energy. But it can't last; before long supplies will begin to run out and alternative sources of energy have to be found, in the form, perhaps, of the hot sulphurous springs favoured by today's thermophilic (heat loving) bacteria that crowd the hot-spots of Iceland or New Zealand. Then came the great invention – capturing the sun's energy, to tap the atmospheric supply of carbon dioxide, to use it to synthesise sugars, and in that chemical process (photosynthesis) to generate and release oxygen, thus slowly but lastingly changing the Earth's atmosphere to the oxygen-rich, carbon-dioxide-poor one we know today (a process that took a billion years or so and which the global burning of fossil fuels is doing its level best to reverse over a few decades). It is here that the great split between primary energy producers – plants – and those that derive their energy by consuming them – animals – emerges.

Enter the nucleic acids

Somewhere over this time must have come what for me is the second great invention, and for many molecular biologists the primary one, that of faithful reproduction. As more material becomes sucked in, so the proto-cells become larger, until, in due course, they become too large for stability and split into two. It seems grandiose to call such division reproduction, of course, and it still lacks the characteristic of all modern reproduction – that is, fidelity. Such proto-daughters are not necessarily identical to their parent, as each daughter cell may contain only a random subset of the molecules present in its parent. Fidelity – faithful reproduction or replication – must have come later.

To generate more or less identical daughter cells means that the same pattern of proteins is present in each. This in turn requires that the mechanism of accurate synthesis of these complex molecules is identical in each cell. This capacity depends on the nucleic acids, which provide the template on which the proteins are synthesised. RNA and DNA, those carriers of the magic letters that constitute the Word, have unique structural properties. The letters form a long string or sequence, of As, Cs, Gs and Ts (RNA has a U instead of DNA's T). When in place along the sequence, the chemical structure of A means that it can also link itself to a separate T, while G can link with C. So a strand of ACGT can serve as a template on which an anti-strand TGCA, can be constructed; and in due course the anti-strand can form the template on which a further ACGT strand can be copied.

$$A - C - G - T$$
$$|\quad|\quad|\quad|$$
$$T - G - C - A$$

RNA exists as a single-stranded molecule, whereas in DNA the two strands, called 'sense' and 'antisense', are locked together to form the famous double helix. As James Watson and Francis Crick recognised fifty years ago, this provides a mechanism for faithful reproduction, for if the DNA strands are separated – unwound – each can form the template on which another antisense or sense strand can be built. If the accurately copied strands of nucleic acid can be utilised by the cell to ensure the synthesis of other key molecules such as proteins, we have established the mechanism now understood as the passage of genetic information from parent to daughter cells. In today's cells this unwinding and copying is subject to multiple and complex controls, and many molecular biologists argue that perhaps in the early proto-cells, the simpler, single-stranded RNA formed the basis for replication. There's a little evidence in favour of this idea, as some modern day viruses use RNA instead of DNA, and, as I've said, some RNAs act as enzymes. But this debate does not concern me here.

There are paradoxes latent in this account of the origins of life, whether replication is based on RNA or DNA. To synthesise a nucleic acid molecule from scratch requires enzymes and energy; and indeed controlled energy production requires enzymes and cellular mechanisms which in today's cells depend for their accurate synthesis on DNA. Turning proto-cells into fully-fledged cellular organisms must have involved a sort of bootstrapping, part of that process of self-creation that some, myself included, call autopoiesis – a term I will unpick more fully in the next chapter. A clue as to how this process may have occurred is provided by the fact that the key molecules involved in cellular energy trafficking are closely related chemically to the molecular building blocks from which the nucleic acids are assembled. But here we are still well within the realm of speculation – though well-informed, I hope.

Evolution – some necessary caveats

Once, however, cells, with their energy supply, enzymes and reasonably faithful mechanisms of replication have been established, we are on firmer ground. As soon as more-or-less faithful replication has evolved, then natural selection begins to work. To say this is not to invoke some magic principle, some *deus ex machina*; natural selection in this sense is a logical necessity, not a theory waiting to be proved.[7] It is inevitable that those cells more efficient at capturing and using energy, and of replicating more faithfully, would survive and their progeny spread; those less efficient would tend to die out, their contents re-absorbed and used by others. Two great evolutionary processes occur simultaneously. The one, beloved by many popular science writers, is about competition, the struggle for existence between rivals. Darwin begins here, and orthodox Darwinians tend both to begin *and* end here. But the second process, less often discussed today, perhaps because less in accord with the spirit of the times, is about co-operation, the teaming up of cells with particular specialisms to work together. For example, one type of cell may evolve a set of enzymes enabling it to metabolise molecules produced as waste material by another. There are many such examples of symbiosis in today's multitudinous world. Think, amongst the most obvious, of the complex relationships we have with the myriad bacteria – largely *Escherichia coli* – that inhabit our own guts, and without whose co-operation in our digestive processes we would be unable to survive. In extreme cases, cells with different specific specialisms may even merge to form a single organism combining both, a process called symbiogenesis.

Symbiogenesis is now believed to have been the origin of mitochondria, the energy-converting structures present in all of today's cells, as well as the photosynthesising chloroplasts present in green plants. Other structures too, such as the whip and oar-like flagellae and cilia that stud the surfaces of many single-celled organisms, enabling them to row themselves around in search of food, and even the microtubules and filaments that provide the cell with an internal skeleton, enabling it to retain its form against external forces, may originally have been derived from organisms with independent existence. This compartmentalisation of function within an individual cell is an important aspect of its regulatory mechanisms, of enabling very complex chemical interconversions to take place within a confined environment. Thus another important early development was to sequester the cell's DNA within an egg-shaped structure, the nucleus, thus helping to keep it under close control. Even today some single-celled organisms – notably bacteria – don't do this, and their DNA is dispersed throughout the cell (prokaryotes). But most larger cells, even free-living single-celled organisms like Amoeba and Paramoecium, and all multicellular organisms, keep their DNA tidly packaged (eukaryotes).

Evolution literally means change over time. To put the account of the emergence of humans and human brains into the evolutionary perspective of this chapter, though, it is important to clear up some common misconceptions.

First, there is no pre-ordained arrow of evolutionary change, some inexorable drive to complexity. There is no tree of life with humans on the topmost branch; no *scala natura*, no higher or lower, no more or less primitive, despite the ease with which these terms are regularly tossed about.[8] All living forms on earth today are there as the consequence of the same 3.5 billion years of evolution, and all are more or less equally fit for the environment and life style they have chosen. I use the word 'chosen' deliberately, for organisms are not merely the passive products of selection; in a very real sense they create their own environments, and in that sense every organism is more or less 'fit' – to use Darwin's own term – for the environment in which it finds itself. Hot springs and volcanic pools only become a relevant environment if an organism with the capacity to survive at high temperature and to exploit the unique chemical environment of the pools evolves to make use of it. *E. coli* adapts its life style to live inside the gut – and humans have evolved to live comfortably with the bug by choosing to make use of its capacity to aid digestion. Water boatmen exploit a property of water that most other organisms ignore and humans are not even normally aware of – its surface tension, enabling them to skim the surface. The grand metaphor of natural selection suffers from its implication that organisms are passive, blown hither and thither by environmental change as opposed to being active players in their own destiny.

Second, natural selection, evolution, cannot predict future change; it is a process only responsive to the here and now. There is no goal, no striving for some metaphysical perfection. Selection can only work on the materials it has available at any time. Evolution works by accretion, by continual tinkering. It cannot redesign and build from scratch. It is, as Richard Dawkins has put it in one of his less contentious similes, like modifying the structure of an aeroplane whilst it is in mid-flight. Furthermore, there are strict constraints on what is or is not possible. These are constraints of structure, of physics, of chemistry. Structural and physical constraints limit the size to which any unicell can grow. Cells need constantly to interchange chemicals with their environment, taking in energy-rich sources, and excreting waste products. The bigger the cell, the harder does this problem become, because its volume increases as the cube of its radius; its surface area only as the square of its radius. So as the cell expands each unit of surface area has to deal with a greater flow across its membrane. This is a structural and physical limit, which is why even my proto-cells eventually split in two as they increase in volume. The chemical constraints are those of carbon chemistry, the variety of possible molecules that can be synthesised, and the extent to which the chemistry of ions, of sulphur, phosphorus and heavy metals can be exploited. The relative biochemical parsimony which means that biochemical processes occurring in bacteria and yeast are very similar to those in humans and oak trees implies that the limits of what could be done by chemical tricks were reached rather early in evolution before these great branches of different life forms diverged.[9]

If terms such as higher and lower are abandoned, how can we describe the evolutionary path that has led from single proto-cells to humans or dolphins or oak trees? One suggestion is that there has been an increase in complexity. But how to measure complexity? The biochemical versatility of many bacteria is greater than that of humans. The Human Genome Project has revealed humans to possess a mere 25,000 or so genes, only 50 per cent greater than the number possessed by fruit flies, and our genome is not only 99 per cent or more identical with that of chimpanzees, which is well known, but about 35 per cent identical to daffodils. If size and biochemical or genetic complexity won't do, the fallback has been another sort of complexity – the number of different cell types (for instance, neurons, muscle cells, red blood cells, etc.) within the organism. On this score humans are said to do quite well, having some 250 or more, an apparent advance on other species.

I will be returning to the questions of the uniqueness of humans repeatedly in later chapters, always remembering that every species is by definition unique. In the rest of this chapter I want to trace the evolutionary path that has led to humans, without entirely forgetting the many branch points on the way. In doing so I will speak of an evolutionary trajectory, from single to multicellular organisms, from nervous systems to brains, from worms to fish, amphibia and mammals, and to do so I will inevitably have to talk about species at present alive, using them, their nervous systems and behaviours as surrogates for those of their ancestral forms along the route. Neither nervous systems nor behaviour leave much by way of fossil traces, although it is possible to make casts of the interior of skulls – endocasts – and draw some conclusion as to brain size, for instance, or to deduce how extinct animals might have moved and what they ate, and hence what their nervous systems must have been capable of. But in general the argument is inferential.

It is always important to remember that describing how brains evolved between fish, amphibia, reptiles and mammals is to imply just that direction to evolution that I have been at pains in the preceding paragraphs to discount. So let me now craftily reinstate it in the language of constraints. Once a particular life style has evolved, there will inevitably be selection pressures to enhance its efficiency. For example: carnivorous mammals feed on herbivores; to do so puts a premium on skills in sensing, outrunning and capturing their prey – which means highly developed visual and olfactory senses, motor skills and the possibility of strategic planning and co-operative hunting. And to survive, herbivores need capacities to sense their predators, the motor skills to attempt to escape them and the social skills of living in herds.

These dramas are played out endlessly on our television screens in innumerable hushed-voice-over natural history programmes. There are selection pressures on each species to develop those skills needed to outwit the other – John Krebs and Richard Dawkins called this bootstrapping an 'evolutionary arms race'.[10] These pressures are to acquire not merely more efficient sense organs and running and climbing skills, but also the brains needed to interpret sense information and instruct motor processes. These constraints lead, in my argument, inevitably towards animals with larger, more adaptive brains, and, I will suggest, ultimately in this planet's life history at least, to the intelligent life forms we call humans. It is in this sense that I may find myself using those old shorthands again of 'primitive' or 'early' or 'higher' life forms. I will try to avoid them, and apologies in advance if you find me slipping into terminologies that are easy but misleading. With these caveats, it is time to return to those earliest of living forms again, and to trace the path by which the past becomes the key to the present.

Living means behaving

By the time that cells capable of metabolism and faithful replication, of symbiogenesis and competition appear, all the defining features of life have emerged: the presence of a semi-permeable boundary separating self from non-self; the ability to metabolise – that is, to extract energy from the environment so as to maintain this self – and to self-repair, at least to a degree, when damaged; and to reproduce copies of this self more or less faithfully. All of these features require something we may term adaptability or behaviour – the capacity to respond to and act upon the environment in such a way as to enhance survival and replication. At its simplest, this behaviour requires neither brains nor nervous systems, albeit a sophisticated set of chemical and structural features. What it does require is the property that some would call a program: at its most general a way of describing both the individual chemical components of the cell and the kinetics of their interactions as the cell or living system persists through time. I'm a bit leery of using the word program, in case it implies that I regard cells as digital mini-computers based on carbon rather than silicon chemistry, but for the present purposes, and provided we recognise that the program is embedded within the cell as a whole, and not in some master molecule within it, I am OK to let it pass.[11]

Built into this program must also be the possibility of modifying its expression, transiently or lastingly, in response to the changing contingencies of the external environment. Environments are inherently variable in both space and time – the technical term is 'patchy'. Thus a free-living cell, perhaps at its largest only half a millimetre in diameter, may find that concentrations of chemical nutrients vary dramatically within a few millimetres or over a few seconds, and it needs to be able to respond appropriately to such changes. One way of conceiving of this capacity to vary a program is as an action plan, an 'internal representation' of the desired goal – at its minimum, that of survival at least until replication is achieved. I will be arguing that, in multicellular organisms, such action plans are ultimately what brains are about.

Amongst the most basic forms of adaptive behaviour drawing on such action plans is goal-directed movement – of a unicell swimming towards food for instance. Dip a thin capillary tube containing a solution of glucose into a drop of bacteria-rich liquid, and the bacteria collect around the mouth of the capillary from which the glucose diffuses – a phenomenon first noted as long ago as the nineteenth century. Such simple responses engage a series of necessary steps. First, the cell needs to be able to sense the food. In the simplest case the food is a source of desirable chemicals – perhaps sugars or amino acids – although it may also be the metabolic waste products excreted by another organism. Indeed the molecule doesn't have to be edible itself provided it can indicate the presence of other molecules that can be metabolised – that is, it acts as a signal. In a watery environment these signalling molecules gradually diffuse away from the source. The diffusion thus provides a gradient – the nearer the food source, the higher the concentration of the signal. But signals are only signals if there is a recipient who can interpret the message they bear. Cell membranes are studded with proteins whose structure is adapted to enable them to trap and bind specific signalling molecules floating past them, and hence read their message. This chemical detection system is the most basic of all sensory mechanisms.

Interpreting the message – using it to develop a plan of action – should make it possible for the cell to determine the direction of the gradient and finally to move up it to the source. Moving towards a specific chemical source – chemotaxis – requires that the cell possess some sort of direction indicator or compass. One way of creating such a compass, employed by bacteria, is to swim in a jerky trajectory, enabling the cell to interpret the gradient by comparing the concentration of the attractant chemical at any moment with that a moment before. That is a time-based, or temporal strategy. By contrast eukaryotic unicells, which are larger, use a spatial strategy, comparing the concentrations of the attractant at different sites along their surface membrane. They can then point themselves towards the highest concentration and begin to move up the gradient.[12]

The molecules trapped by the receptor on the surface membrane serve as signals, but very weak ones. To produce as dramatic a cellular response as turning and moving in the right direction requires that the signals are highly amplified. The mechanism by which this is carried out, even in the seemingly simplest of unicells, turns out to be the basis on which the entire complex apparatus of nervous systems and brains is subsequently built. The receptors are large proteins, oriented across the lipid membrane, with regions sticking out into the external environment, and also 'tails' which reach into the interior of the cell (the cytoplasm). When the signal molecule binds to the receptor protein its effect is to force a change – a twist, if you like – in the complex shape of the receptor. This twist is sufficient to make the membrane temporarily leaky, allowing ions such as those of sodium or calcium to enter. In turn these can trigger a further cascade of chemical reactions within the cell. This cascade culminates in the cell turning and orienting towards the food source. What determines the response is not some great change in the total amount of calcium in the cell but a pulse, entering and spreading rapidly through the cell like a wave within a few thousandths of a second. This mechanism, presumably developing very early in evolutionary history, is conserved and put to use in much more recently evolved nervous systems too.

The end result is that the cell begins to move up the gradient. Some unicells, like Amoeba, will ooze, extending their cell membrane in the desired direction and retracting it in the rear, like a miniature slug. This oozing involves a set of protein filaments that provide both an internal cellular skeleton, and internal 'muscles'. Indeed the analogy is apt because one of the key protein filaments is actin, which is also one of the two major proteins constituting muscle (the other is called myosin). A unicell, like Paramoecium, which eats by engulfing bacteria, moves by rowing through the surrounding medium with its array of cilia – a process also involving actin. The cilia must beat in co-ordination, achieved by a system of fine protein threads running longitudinally, connecting the cilia at their bases. Except when it is actually feeding a Paramoecium is constantly moving around, until it finds a food-rich area. If by chance it then leaves this self-service restaurant, it moves back into it again by reversing the beat of the cilia on one side. It also goes into reverse, just like a battery-driven toy motorcar, if it bumps into an obstacle in its path. It will similarly avoid sources of excessive heat or cold, or of irritating chemicals such as sulphuric acid. One way of speaking of this process, favoured by neurologist Antonio Damasio, even in so limited an animal as Paramoecium, is as 'expressing an emotion'. Emotion, for Damasio, is a fundamental aspect of existence and a major driver of evolution.[13] I'll have more to say about this later.

Meanwhile, it is important to recognise that these principles of converting sense data into planned movement are general, and don't apply just to chemical attractants. For example, unicells such as Euglena, that obtain energy by photosynthesising, and which contain large red granules of light-sensitive chemicals, are phototropic – that is, their action plan ensures that they move towards the light, maximising the amount of the sun's energy that they can trap.

Genetic technologies have made it possible to study some of these processes in greater detail. It is, for example, possible either to delete entirely, or temporarily inactivate, specific genes within a cell, enabling one to engineer bacteria which lack the surface receptor molecules; all other parts of the system are intact, but because the cells cannot detect the chemical gradient, they have no way of directing their movement, though they will still willingly absorb and metabolise the molecules if placed amongst them. In this way the individual steps along the chemo-tactic pathway can be dissected out.

Multicellularity and multiple signalling systems

It is probable that for most of life's history on earth, the planet was populated only by single-celled organisms. At some point however, the crucial discovery would have been made that there were benefits to be had from cells teaming together into larger aggregates. At first, this teaming up would have been in the form of temporary coalitions. Slime moulds are a good example. For part of their life cycle they get on well as independently-living amoeboid cells, but at other crucial phases – for instance in periods where food is in short supply – the individual cells come together to form a multicellular mass – another example of the importance of co-operative processes in evolution.

Temporary or permanent, multicellularity makes for a profound change in life style. For free-living cells, 'the environment' consists of the world outside them, and each has to have the capacity to respond adaptively to rapid changes in this patchy world. The merit of social living is that it is no longer necessary for each individual cell to retain all these mechanisms, but this also means that the survival of any individual cell is dependent on the survival of the organism as a whole, and each must surrender its independence to help preserve that community

– to communicate in order to co-operate. Evolutionary mechanisms are highly conservative, and there are remarkable similarities between the overall set of biochemical processes occurring within unicells and those performing the routine housekeeping functions within individual cells in a multicellular community or fully-fledged organism. Within a multi-cellular environment cells can specialise, each type maturing into a different shape and requiring a different subset of proteins to be synthe-sised over and above each cell's basic housekeeping needs. Specialised sense cells pick up the messages impinging from the external environ-ment, and other specialised contractile cells (the forerunners of muscles) provide the mechanisms of motility. Some cells in the society continue to face the external world, but others – most – live their entire lives deep inside the body, and for these their environment is no longer patchy but maintained within fairly close limits by regulatory mechanisms oper-ating at the level of the organism as a whole.

This is what the nineteenth-century French physiologist Claude Bernard described as 'the constancy of the internal environment' – one of the most famous slogans in the history of biological science.[14] These days this is referred to as homeostasis, but, for reasons outlined in my book *Lifelines*, I prefer to use the term *homeodynamics*,* to emphasise that stability is achieved not through stasis but through dynamics. It is fascin-ating to discover that many of these regulatory mechanisms involve internal signalling processes that, biochemically, are variations on a theme already invented by the unicells, but now adapted for the condi-tions of multicellularity. This take-over of structures and processes orig-inally evolved as an adaptation to one function for use in a totally different way, is another constant feature of evolution through life's history, a process called exaptation by the late great evolutionist Stephen Jay Gould.[15]

With multicellularity, 'behaviour' becomes a property of the organism as a whole, to which the 'needs' of individual cells are subordinated. The internal representation which makes possible the action plan for the organism can be delegated to specific cell ensembles. This requires new modes of communication to be developed. Where previously there were only two classes of signals – those arriving from the external envi-ronment to the cell surface, and those internal to the cell – there are

---

* Others have used the term homeorrhesis to refer to the same phenomenon.

now three. Signals from the external environment are still registered by sensory cells on the surface and are transmuted by molecular cascades within them, but now the response to these cascades requires that further messages be sent from the sensory cells to other regions of the body, including of course the contractile cells. Sometimes the sensory cells make contact with intermediaries whose task it is to synthesise and secrete the necessary 'messenger molecules'. The messengers can then be distributed through the body either by way of a circulatory system or by diffusion through the extracellular space between the body cells, and are detected as before by specialised receptor proteins on the surface membranes of their targets. When molecules that served such messenger functions were first identified in mammals, they were given the generic name of hormones. It was only later, and to some surprise, that it was discovered that many of the same molecules also serve as intercellular signals in very early multicellular organisms, another powerful example of evolutionary conservation.

Intracellular signals are fast; calcium waves may cross a cell within milliseconds; but the timescale on which the extra-cellular messengers act is inevitably slower, being governed by rates of diffusion between cells and through circulatory systems such as blood or sap. Furthermore, although the message may be intended only for a particular cell type that has the receptors capable of responding to it, it is not a smart signal, capable of precise direction; enough must be produced to reach all body parts so as to ensure that some at least gets to the target.

All multicellular forms of life, plants and fungi as well as animals, employ such messengers. It is to the divergence between these great kingdoms that we must look to find the origins of nervous systems as such. Plants are largely immobile; they obtain their energy by photosynthesising, and to do this it is sufficient for them to spread their leaves and grow towards the light and, if necessary, to turn their heads in response, as do sunflowers. For this, relatively slow intercellular signalling processes are sufficient.*

* Of course, plants do make use of inter-organism signals as well. Scents that attract pollinating insects are obvious; less obvious are the volatile chemicals released when the plant is stressed, for instance when attacked by pests, and which may serve to deter further assaults either on the injured plant or its neighbours. Such airborne signals – pheromones – may be effective over long distances and are even more time-relaxed than are the hormones.

However, organisms that depend on finding other, presynthesised forms of food to provide energy – that is, animals that prey on plants, to say nothing of animals that prey on other animals – need to be able to move, and to co-ordinate their movements and the activities of their varied body parts more rapidly and precisely than is possible simply by spreading a general message throughout the body. It is more efficient, indeed necessary, to be able to have a direct line of communication between sensory and effector cells – direct dialling private lines. It is this that a nervous system is able to provide (Fig. 2.2).

It's easy to imagine a sequence whereby neurons evolved from secretory cells. Instead of discharging their contents generally into the surrounding space and circulatory system, the secretory cells could have put out feelers (called 'processes') enabling them to make direct contact with their targets, so as to signal rapidly to them and them alone. Messages could be conveyed between the two either electrically or chemically – by a depolarising wave or by secreting a messenger molecule across the membrane at the point where the two cells touch. In fact both phenomena are known to occur.

Nerve nets and nerve cells

The first step towards such nervous systems can be seen among the large group of Coelenterates, believed to be amongst the earliest true multicellular animals. The best known is perhaps the Hydra, a tiny creature that sits at the bottom of streams attached to rocks or water plants, waving its tentacles above its mouth. When a potential source of food brushes past its tentacles, the Hydra shoots out poisonous threads, collects the paralysed victim and thrusts it into its mouth. Like the sea anemone, Hydra closes down to a blob if touched. A well-fed Hydra is quiescent; when hungry it waves its tentacles vigorously, or moves its location by repeatedly turning head-over-heels, seeking food-rich or oxygen-rich environments (once again, Damasio would regard these acts as 'expressing emotions'). These quite complex forms of behaviour demand numbers of different specialised cell types: sense receptors responsive to chemical or touch signals, secretory cells, muscle cells, and also primitive nerve cells, all embedded in sheets of tissue composed of the cells ('epithelial') that make up the bulk of the body mass. There

Before nervous systems

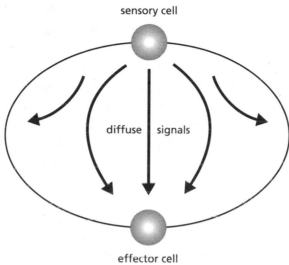

With a nervous system

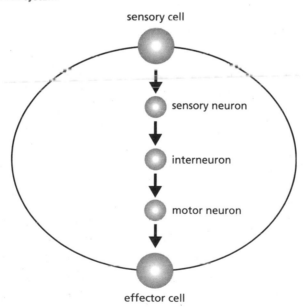

Fig. 2.2 *Constructing a nervous system.*

needs to be a high degree of co-ordinated activity amongst these cells to prevent mistakes being made – for instance to distinguish between something which is a source of food or something which is potentially dangerous, and to respond appropriately by firing poison or closing down. The mouth must be open at the right time and the gut muscles controlled or the Hydra will swallow its own tentacles – though fortunately for its survival it cannot digest its own cells.

Hydra co-ordinates its cells partly by electrical signals – bursts of activity, maintained in part by calcium gradients. In response, the body contracts in a wave starting at the base of the tentacles and spreading at a rate of about fifteen centimetres a second. In addition there is a steady slow electrical beat arising from various points on the body surface which can quickly change its frequency in response to environmental change, as when a Hydra which has been in the dark is exposed to light. As well, however, scattered through the Hydra's body there is a fine network of neurons – a conducting system linking sensory and effector cells forming the beginnings of a true nervous system.

Sense receptors respond to the external environment, effector cells act upon it; but integrating information arriving from a variety of sense receptors, and summing their output in the form of instructions to perhaps many different effector cells, requires an intercommunicating network of cells. These are the neurons, which thus provide the principal component of the organism's action plan.

As nervous systems become more complex, there are many intervening stages as networks of interconnected neurons collate information in the form of signals arriving from multiple sources before integrating them and then distributing them to multiple effector organs. Many of these interconnecting neurons (interneurons) do not have direct connections with either sensory or effector cells, but interact richly one with another through multiple feedforward and feedback loops (Fig. 2.3). It is within this network of inter-neurons that the organism builds up its internal models of the external world and co-ordinates its plans of action on that external world. If an essential feature of living organisms is the possession of an internal program, then in multicellular animals it is the neural net – the system – rather than the individual cell that embodies that program.

A typical neuron, the unit within this network (Fig 2.4), has, like all cells, a cell body containing enzymes and structures necessary for

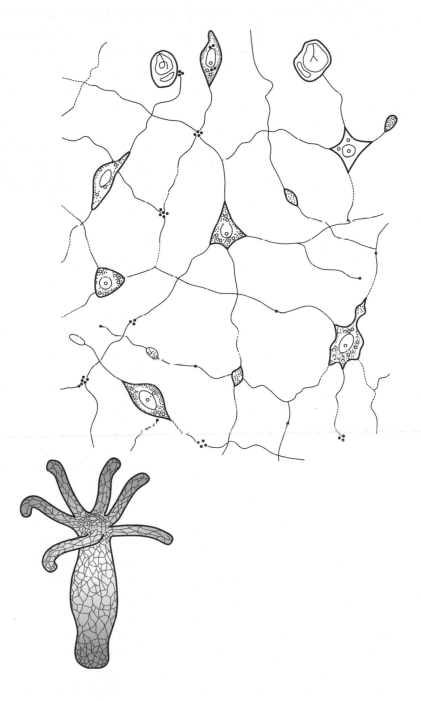

Fig 2.3 *Hydra and its nerve net.*

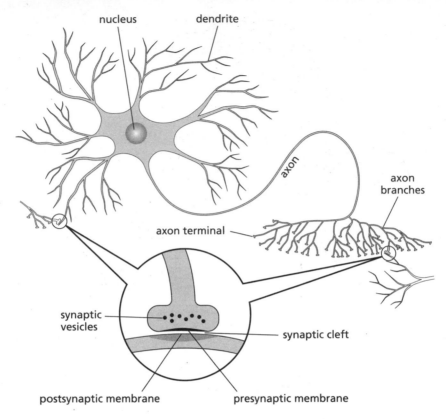

Fig 2.4 *The neuronal archetype.*

biochemical housekeeping mechanisms, such as the DNA-packed nucleus, mitochondria for energy production, and so forth. However, spreading out from the cell body is a tree-like structure of branching processes, called dendrites, and a long thin tail, the axon. The dendrites are the collecting points; it is there that other cells, sensory cells or other neurons, make contact and transmit messages forward chemically or electrically. The axon is the conducting path down which the integrated signal, summed from the activity in the dendrites, passes to another neuron or to an effector cell.

The junctions at which one cell signals to another are known as synapses and they are central to the functioning of the system (Fig. 2.5). Plant cells can signal across their membranes by changes in electrical potential, and for many years after synapses were discovered and named by the neurophysiologist Charles Sherrington at the beginning of the

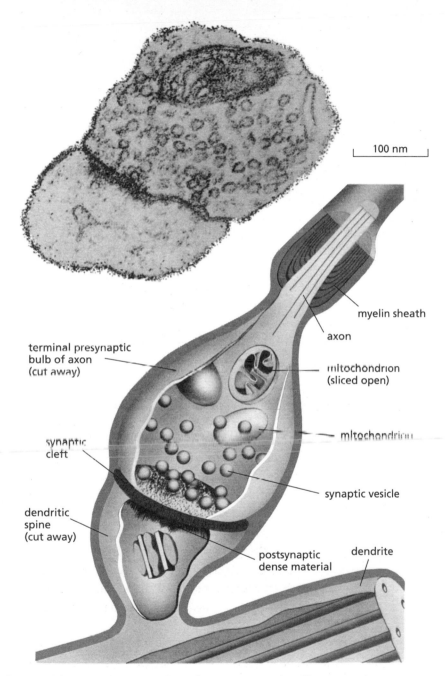

100 nm

myelin sheath

axon

terminal presynaptic
bulb of axon
(cut away)

mitochondrion
(sliced open)

mitochondrion

synaptic
cleft

synaptic vesicle

dendritic
spine
(cut away)

postsynaptic
dense material

dendrite

Fig 2.5 *Electron microscope view of a synapse, and a diagrammatic reconstruction. The human brain contains some 100 trillion synapses. Note scale bar.*
1mm = 10⁻⁹m.

last century (in a term derived from the Latin for 'to clasp'), transmission in the brain was also thought to be primarily electrical. Only by the end of the 1930s did it become clear that most, though not all, such signalling was chemical. Many of the chemicals that serve to carry signals between neurons – known as neurotransmitters – are derived from molecules already present and with other functions elsewhere in organisms without nervous systems. The neurotransmitter noradrenaline (norepinephrine in the US) is related to the hormone adrenaline, produced by the adrenal glands. Another neurotransmitter, serotonin, serves elsewhere as a regulator of cell division.

A synapse is a junction: the transmitting side arising from the axon in a bulge in the membrane, behind which are stacked an array of tiny spheres – vesicles – packed with neurotransmitter. When a signal arrives down the axon from the nerve cell body itself, the vesicles move to the membrane, releasing the neurotransmitter into the gap between the transmitting and receiving cell. The reception points are on the dendrites of the receiving neuron (or directly on to a muscle) and they consist of specialised regions of the membrane containing protein receptors to which the transmitter can bind, causing a cascade of biochemical and electrical activity in the post-synaptic cell. Neurons can make and receive many synapses – up to several tens of thousand per cell in the human brain, though far fewer in the much simpler nervous systems that characterise Hydra.

What distinguishes a fully-fledged nervous system – our own for instance – is a one-way flow of information through the system, from dendrites to axon, from sensory cell to effector. Of course this is mediated via all the feedback loops, but none the less there is a directionality to it that the Hydra's nerve net does not possess.

Whereas Hydra's neurons are scattered throughout its body, the next crucial step was to concentrate them within an organised system. A spadeful of soil dug from almost any environment on earth is packed with living creatures – myriad bacteria and unicells of course, but also tiny worms called nematodes. There are anywhere between 100,000 and ten million different species of nematode, ranging from about half a millimetre to as much as a metre in length. One, *Caenorhabditis elegans*, now bids fair to be the most studied animal on the planet. For the past thirty years, teams of researchers have pored over the most intimate details of its life cycle, the patterns made by its precisely 959 body cells, its sex life and eating habits (it is a hermaphrodite and feeds on bacteria), and not least, its nervous system, consisting of exactly 302 cells, each one of which has a specific role. *C. elegans* has a head and a tail end, and as it is more important for it to know where it is going than where it has been, many of its sensory cells are clustered at its head end. From these, nerve connections run to clusters of interneurons, packed into groups (ganglia) with short interconnecting processes between the cells within the group and longer nerve tracts leading out along its gut and ultimately to the effectors: contractile, egg- and sperm-producing cells. These neurons use many of the neurotransmitters that are found in mammalian brains (notably the amino acid glutamate), an indication of how far back in evolutionary terms these molecules were adapted for signalling functions.

The functional wiring diagram of these nerve cells, virtually identical from individual worm to worm, has been mapped in detail, as has their pattern of development during their rapid maturation from egg to adult (so fast do they breed that a single hermaphrodite can produce 100,000 descendants within ten days). It also becomes possible to relate the pattern of nerve cells to specific forms of behaviour. *C. elegans* moves in a characteristically sinuous, wormy way – but, because its pattern of nervous connections is so precise, if specific cells are removed (for instance by laser ablation) or connections cut, its pattern of movement changes. Mutant worms can be generated which can only move forwards and not backwards, others that swim in unco-ordinated jerks, others whose sexual behaviour becomes abnormal. The worm has become a toy model for geneticists, developmental biologists, neuroscientists.[16] The complexity of behaviour possible within this minuscule organism and its limited nervous system is extraordinary. Its repertoire continues to produce surprises. For instance some *C. elegans* are solitary, others seek company, engaging in synchronised swimming and social feeding on their bacterial prey. (Another example of co-operative behaviour; I continue to emphasise these phenomena in part to counter the myth that evolution and Darwinian natural selection are all about competition. There is nothing anti-Darwinian about this; natural selection can, and under the right circumstances does, favour co-operation.) It turns out that social feeding is induced by two specialised sensory neurons that detect adverse or stressful conditions such as noxious chemicals in the environment; ablation of these neurons transforms social animals into solitary feeders.[17]

If so much can be achieved with such a small number of cells and interconnections, it should be no surprise that, as the number of cells increases in larger organisms, so does the complexity and variety of possible behaviours. Nervous systems provide programs and action plans for responses to the environment, but it is important that these plans are flexible and can be modified by experience, and nervous systems have this capacity built in at an early stage. Put a piece of raw meat into a stream and within a few hours it will be covered by little flat black worms, feeding in it – Planaria, about a centimetre long. Planaria have light sensitive cells embedded in pits at their head end, avoid light, and respond to touch and chemical gradients. If a planarian is touched with a glass rod it will curl itself up into a ball, its response to danger. Slowly and cautiously it will uncurl again. Repeat the touch, and it will repeat the response. However, if the touch is repeated enough times, the response will diminish and eventually the planarian will no longer respond – it has become accustomed to the stimulus of the rod and no longer regards it as dangerous. This process is called habituation, and is a universal property of nervous systems. It could be regarded as a primitive form of learning, and its biochemical mechanisms have been extensively studied.

We all experience such habituation many times every day, starting from the moment we dress in the morning. We start by being very aware of the feel of cloth against skin, but within a short time cease to notice the feel. However, if the nature of the stimulus is slightly modified, our clothes unexpectedly rub up against something – or, for the planarian, the touch of the rod is coupled with a squirt of water – then the full response will immediately reappear. We are aware once more of our clothes; the planarian senses danger and curls up into a ball again. Habituation does not occur simply because the animal is tired or some chemical mechanism is exhausted, but is a way of adapting based on experience. Neither is the habituation a permanent change in behaviour; given a long enough gap in time the original response reappears.

Ganglionic brains

The evolutionary track I have been mapping has led from proto-cells to faithfully replicating eukaryotes capable of responding adaptively to patchy environments; from single-celled eukaryotes to multicellular animals with internal signalling systems, and from these to fully-fledged nervous systems capable not merely of constructing action plans, but of modifying those plans, at least temporarily, in response to environmental contingencies. But we haven't yet arrived at brains. This must have been the next step along the evolutionary path that led to humans. Concentrating neurons in ganglia is a way of enhancing their interactions and hence their collective power to analyse and respond to incoming stimuli. Locating them at the front end of the organism is the beginning of establishing not merely a nervous system but a brain, though head ganglia or brains only slowly begin to exert their primacy over the other ganglia distributed through the body. These remain capable of independent action. Cut off the abdomen of a wasp, and the head end goes on feeding even though the food can no longer be digested. Even in the octopus, the individual ganglia retain a degree of independence. If the octopus loses one of its eight tentacles, the severed arm will go on threshing about for several hours.

There is, let me emphasise once more, no linear evolutionary track from proto-cell to humans; rather there are multiple divergent paths. The overwhelming majority of species alive today manage without brains or even nervous systems – and they manage very well; they are no less and no more highly evolved than are we. Present-day life's rich diversity is a web of mutually interacting organisms that have evolved vastly different ways of making a living. Evolving life is full of branching points. Mention of wasps and the octopus – a mollusc – brings us to yet another set of branch points in the history of the brain. For although insect (arthropod) and molluscan neurons are pretty similar to human neurons, and the biochemical motors that drive the system – their electrically excitable membranes and the neurotransmitters – work in the same way, the organisation of the system is entirely different. In molluscs and arthropods the central ganglion – the nearest any amongst these huge numbers of species have to a brain – and the principal connecting pathways between it and other ganglia lie arranged in a ring around their guts. This is a device that can be seen even in earthworms, and it imposes a fundamental design limitation on the complexity of the nervous system. As the number of neurons increases so the nerve ring round the gut must thicken, and this tends to reduce the diameter of the gut itself. This limitation is sharply revealed, for instance, in spiders, whose guts are so narrowed by their nerve rings that they can only digest their prey as a thin trickle of liquid. The problem of disentangling guts from brains prevents any dramatic increase in the size and complexity of the nervous system – a blockage made even sharper in arthropods, which have no internal skeleton but instead have hard external shells whose structure gives little scope for growth. The best that can be done is to distribute the 'brain' into a number of discrete spatially separated lobes, each with separate functions, but linked by tracts carrying hundreds of thousands of individual nerves ensuring efficient communication.

Limits provide opportunities. The myriad life forms of insects and their complex behaviours show that big brains are not necessary for evolutionary success. Many insects have extremely well-developed sensory systems. These include eyes built on a totally different principle from vertebrate eyes and sometimes with the capacity to distinguish different wavelengths of light – that is, colour vision; and a keen sense of smell that enables them to detect the plant or animal odours that indicate prey. Anyone who has ever been subject to assault from mosquitoes can attest to some of these skills. However, what the limitations on brain size do mean is that much of this behaviour is hard-wired – fixed within the pattern of connections – and not capable of much modification in response to experience. A fly trapped on the wrong side of a half-open window cannot learn to fly round the obstacle, instead stubbornly banging its head on the glass as it follows its attempt to navigate towards the light. But interrogate them ingeniously enough and you will find that even flies can learn some skills. Fruitflies – Drosophila – can discriminate between different odours, and if they are given an electric shock as they fly towards one particular source of scent, they will tend to avoid it subsequently.[18] And as with *C. elegans*, mutant flies can be bred, or 'constructed' by genetic manipulation,* that lack particular skills. For instance, deletion of the genes responsible for coding for particular proteins can result in flies that cannot learn, or can learn but not remember. Such techniques have helped elucidate some of the biochemical processes required for learning and memory formation. However, the sheer number of neurons and complexity of the networks means that the exact pathways involved within the various ensembles

---

* The resonances of this term 'constructed', for genetically manipulated plants and animals, are worth pondering.

of neurons at the fly's head cannot be as easily mapped as is possible in *C. elegans*, for instance.

Social insects, such as ants and bees, carry these skills to almost legendary heights, necessary in their complex societies composed of genetically identical individuals. They can signal to one another in complex and meaningful ways. The entire ant colony has to be involved in choosing the best nest site, and to do so must assess information caming in from different sources, brought back by its scouts, and then arrive at a collective decision. Some people have even referred to the colony as a super-organism that shares a collective intelligence. But even individual skills can be formidable. Ants can navigate away from their nests distances equivalent to hundreds of kilometres for humans, and use a variety of compass-like techniques to steer their way back.[19] They can distinguish their own from another colony's path by sensing specific pheromones, signal the presence of food or enemies, and even produce 'propaganda' pheromones to disorient potential prey.

Honey bees can do much more, navigating from their hives, discriminating potential flower food sources by scent and colour, returning to their hives using the sun to provide compass directions and then performing their famous waggle dances to inform their colleagues of the nature and distance of the food sources. They readily learn new patterns of behaviour as one flower source becomes exhausted and another must replace it. In the laboratory they can be taught to respond to novel scents by extending their proboscis to drink from sugar solutions. The cells involved in learning these skills, the pathways between them, and some of the biochemical mechanisms involved have been studied in great detail.[20]

Like insects, molluscs too have brains distributed across several ganglia. The largest and most complex molluscan brain is that of the octopus, whose several lobes contain as many neurons as does the rat brain. Octopodes can learn quite complex tasks, based on touch via their tentacles, and on sight. They can be taught to discriminate between rough and smooth, black and white, square and circle. (One mollusc, the large sea slug *Aplysia californica*, which is characterised by having very large neurons identifiable from animal to animal, rather like *C. elegans*, even earned the man who exploited the study of its learning ability a Nobel Prize.[21]) However, skilled as these modern animals are, the evolutionary route by which they emerged diverged from that which led to large vertebrate and ultimately human brains long, long ago.

Real brains at last

The development of large brains required two major changes in the construction of nervous systems: the separation of the nerves themselves from the gut, and the concentration of nervous power. It also required the first step towards the development of a bony skeleton. Amphioxus, a small sea-floor fish, is an example. Less behaviourally sophisticated than octopus or bee, it has a flexible rod of cartilage, a notochord, running down its back – the forerunner of the spinal column – with the merit of providing a bracing device against which muscles can pull. More relevantly for the present argument is that the major nerves and central ganglion also lie in a continuous tube running the length of the creature's body, thus disentangling them from the gut and giving space for growth.

The next design improvement was the invention of bones rather than cartilage to provide an internal framework for the body, unlike the arthropods' exoskeletons. The neural tube could then be encased within the spine, thus protecting it, but leaving the head end, with its concentration of sense receptors, free to expand; early vertebrates already have three large neuron-packed swellings at the head end of the neural tube, comprising hind-, mid- and forebrain, each associated with one of the special senses. The forebrain registers smell, the midbrain vision and the hindbrain equilibrium and vibration (Fig 2.6). This description implies a division of functional labour between different brain regions, and serves to emphasise that it is perhaps a mistake to speak of 'the brain' in the singular. In reality it is an assemblage of interacting but functionally specialised modules – the descendants of ganglia. It is a plural organ – but one that normally works in an integrated manner. Just how this is achieved, how *e pluribus unum* becomes possible, in such contrast to the semi-independent ganglia that comprise insect nervous systems, is a central question that will come to haunt later chapters.

This is the *bauplan* (architect's plan) on which all subsequent vertebrate brains are built. Later-evolving brains show an increase in size, and hence neuronal number; the forebrain becomes the cerebrum, itself

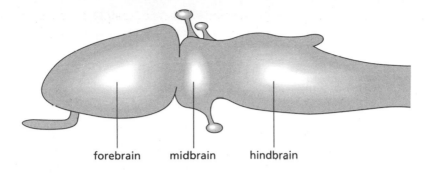

forebrain   midbrain   hindbrain

Fig. 2.6 *The* bauplan *of the vertebrate brain.*

divided into a forward region (telencephalon) and a rear (thalamen-cephalon); the midbrain becomes the optic tectum; and the hindbrain becomes the cerebellum. In bony fish, the chemical sense receptors – now better regarded as smell detectors – form olfactory bulbs in the nose, from whence nerves run to the cerebrum. Fish also have well developed vision, and the light sensitive eye spots or eye pits of Planaria and Amphioxus have become transparent windows containing lenses, with light-detecting cells forming the retina, itself an outgrowth of the neural tube, beneath them. Such an eye does more than merely detect patterns of light and dark; its lens enables an image of the external world to be presented to the retina and thence to the brain proper. To handle the more complex information that such a refined system provides, the optic tectum expands, forming two separate lobes, packed with neurons.

This increase in size, however, presents another design problem. The more neurons, the more complex the pattern of interconnections between them. How to pack them all in and keep the wires from becoming crossed? One device is to coat each axon in an insulating layer of lipid, rather like the plastic sheath on electrical wiring. The lipid is called myelin; white in colour and rather greasy in texture, it forms the brain's well-known 'white matter' as opposed to the 'grey matter' of densely packed neuronal cell bodies and dendrites (actually in real life rich with blood and therefore pinkish rather than grey). Myelin is synthesised by specialised cells which form part of the nervous system but are distinct from neurons. Called glia, they come in a variety of flavours. As well as those that make myelin, others surround the neurons within the grey matter, ferrying foodstuffs to them from the blood supply and helping dispose of waste material. Glia greatly outnumber neurons in the mammalian brain and, as will become clear in the next chapter, also have a crucial role in brain development.

This insulating system was invented fairly early on in evolution. In the Amphioxus neural tube, as in the human spinal cord, the grey neuronal mass is in the centre, wrapped around by the connecting axons in their myelin sheaths. As with the arthropod nerve rings round the gut, this is a size-limiting design. A better approach to the packing problem, found for the first time in the fish's optic lobes, is to put the white matter inside, and surround it with a thin skin, or cortex, of grey matter, a few layers of neurons deep. This helps solve the problem – a small increase in surface area of the cortex can make a very large difference to cell numbers without greatly increasing the overall volume of the lobe. Not all brain modules are organised in this cortical manner, either in fishes or other vertebrates, but it is a solution adopted as relative brain size and complexity increases in mammals, primates and humans. Eventually even this is not sufficient, and to increase cortical area still further the surface of the cerebrum becomes wrinkled, with deep valleys (sulci) and hillocks (gyri). Humans and other primates, and dolphins, have intensely wrinkled cerebral cortices.

Subsequent evolutionary transitions take the form both of a steady increase in brain size relative to body mass, and also a subtle shift in functions. It is with the first land-based animals, the amphibia and reptiles, that the great development of the forebrain region begins. Perhaps this may be associated with the fact that a sense of smell is a good deal more important for a land-based than a marine animal. Whatever may be the reason, in amphibia, reptiles and birds the forebrain is enlarged at the expense of the optic tectum. While some visual and auditory inputs are still mediated by way of these midbrain regions, the major processing of both visual and auditory stimuli has now moved irrevocably forward. Connections run from the optic tectum up to regions in the forebrain which enable an additional level of visual analysis and decision-making to occur, and the functional independence of the tectum is correspondingly reduced. The thalamus (part of the thalamencephalon) begins to assume a major co-ordinating role.

Even though to the naked eye the forebrain looks like a mass of homogeneous tissue, under the microscope it can be seen to be divided into numerous distinct regions, each packed with neurons. The anatomical divisions between them are barely visible, but their distinct functions are increasingly understood. Thus in the bird brain there are discrete regions densely packed with neurons associated with detection of taste, odour and sound, with navigation and spatial memory and, in songbirds, song learning, as well as modules organising and controlling the outputs of behaviour such as feeding or song production.

Mammalian brains

The evolutionary development from the amphibians through the reptiles to mammals results in the dominance of the furthest forward part of the brain, the telencephalon, which in mammals developed from the olfactory lobes so as to swell outwards, enlarging and folding over all other brain regions to form the cerebral hemispheres. With the mammals the cerebrum takes over the task of co-ordination and control from the thalamus. Some of the thalamic regions become mere staging posts, relay stations en route for the cerebral cortex. Some, however, such as the hypothalamus and pituitary, remain of vital significance in control of mood, emotion and complex behavioural patterns. The hypothalamus contains groups of neurons concerned with the regulation of appetite, sexual drive, sleep and pleasure; the pituitary regulates the production of many key hormones and forms the major link between nervous and hormonal control systems. Those who would stress humans' relationship with other animal species always point out how critical these drives and behavioural states are for humans, how much they dominate the totality of human behaviour and what a large proportion of total human existence is taken up with activities associated with or driven by them. Humans have in the core of their brain, such popularising behavioural determinists have maintained, a 'fish brain' and a 'reptilian brain' which are in many ways more important than the much-vaunted cerebral cortex. It is true that in the evolutionary development of the brain, few structures have ever been totally discarded. Rather, as new ones have developed the old ones have become reduced in importance and relative size, but many of the connections and pathways remain. It is also true that the hypothalamus is of considerable importance in mood and behaviour determination, in mammals and even humans.

However, to extrapolate from these facts towards the claim that because similar brain structures exist in people and frogs, people's behaviour is inevitably froglike, is nonsense. It is like arguing that we think by smelling because the cerebral hemispheres evolved from the olfactory lobes. Brain regions remain, but their functions are transformed or partially superseded by others. Fish, amphibia, reptiles and birds survive today because they are fully 'fit' for their environmental and life styles – at least as fit and as 'evolved' as humans are. The processes of evolution by which their ancestors emerged ensured that. The evolutionary line that led to mammals and thence to primates and humans describes but one of the myriad evolutionary trajectories that have generated all current living forms and that continue to drive adaptation in response to ever-changing environmental contingencies. Above all, adaptation is about survival; our brains evolved as a strategy for survival, not to solve abstract cognitive puzzles, do crosswords or play chess.

Even amongst the mammals themselves, the changes which the cerebral hemispheres undergo are considerable. Monotremes or marsupials like the Australian duck-billed platypus or the North American opossum have well-developed forebrains by comparison with the reptiles, whose cerebral hemispheres have a cortex not more than a single layer of cells thick. By contrast even the early mammals have a cortex with many layers of cells. In the early mammals the cortex is probably almost entirely olfactory in function, but the forward movement of control continues, with the integration of information from different sensory modalities dependent more and more on cerebral cortical dominance over thalamic control. The greatest expansion occurs with the formation of the neocortex (Fig. 2.7). As this increases in area, so the older cortical regions are forced deeper into the brain structure, becoming curved in the process to form the region known in mammals as the hippocampus, which has a central role in the formation of memory, and especially spatial memory (the hippocampus has famously been called 'a cognitive map'[22]). An index of this development is the fact that where in mammals like the hedgehog the ratio of the volume of the neocortex to hippocampus is 3:2, in monkeys, it has increased to about 30:1.

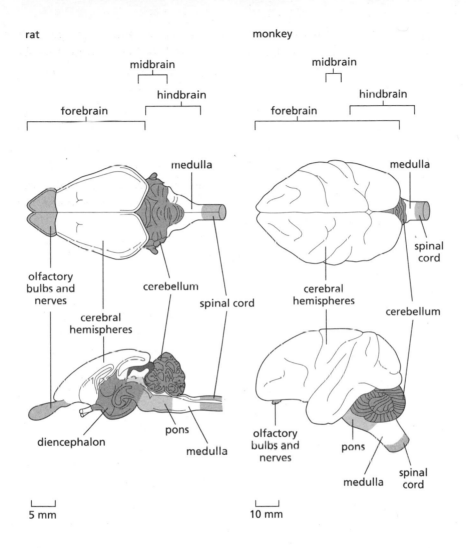

Fig 2.7 *Key structures in the rat and monkey brains (note differences in scale!).*

The cortex takes control

All the cerebral regions except the neocortex have some sort of rudimentary equivalent in reptiles; it is the layered neocortex which is unique to mammals, and the way in which it has taken over thalamic functions can be shown by mapping the connections between neocortex and thalamus, which all arrive within specific layers of the cortical neurons: each thalamic region is found to match an appropriate neocortical area. In the earliest mammals, as in today's marsupials, the motor area and the cortical targets of thalamic neurons occupy most of the neocortex. Functionally, then, the neocortex must be largely concerned with the more sophisticated analysis of information which in the amphibia is handled by the thalamus alone. The major development in the later-evolving mammals is the expansion of the area of neocortex between the sensory and motor regions. These contain neuronal ensembles (sometimes called association areas) which do not have direct connections outside the cortex, but instead talk only to one another and to other cortical neurons; they relate to the outside world only after several stages of neuronal mediation. In humans, these areas include the massive prefrontal lobe and regions of the occipital, temporal and parietal lobes.

Functionally, these expanded association areas are clearly *acting upon* information which has already received quite sophisticated analysis, with integrative areas processing inputs from multiple sensory systems and relating them to prior experience. Just as the brain as a whole is not a single organ but an accretion of less- and more-recently evolved structures organised into distinct modules, so too is the cortex. The cerebral cortex, some 4mm thick in humans, contains about half the neurons in the entire brain, arranged like a layer cake in six tiers; the pattern is readily observable in side view through the light microscope if the cells are appropriately stained. Less observable is the fact that, viewed from its upper surface, the cortical neurons are also organised into an array of functionally distinct columns, running at right angles to the surface of the brain with sharp boundaries between them. Closer examination of neurons reveals that each has a different but specific shape classifiable as pyramidal, stellate (star-shaped) and basket (Fig. 2.8). Each cortical layer contains a range of different neuronal types, pyramidal cells being the most common. Different-shaped neurons have different patterns of connectivity and therefore are functionally distinct. Each neuron is connected to others, both to some of its neighbours and to some distant from it, via its dendrites and axons. In particular it is the pyramidal cells (some 75 per cent of the total) that send their axons over the longest distances, the other forms being interneurons, making only local connections.

A good example of this complexity is provided by the visual system, one of the best mapped of all brain regions. In primates visual

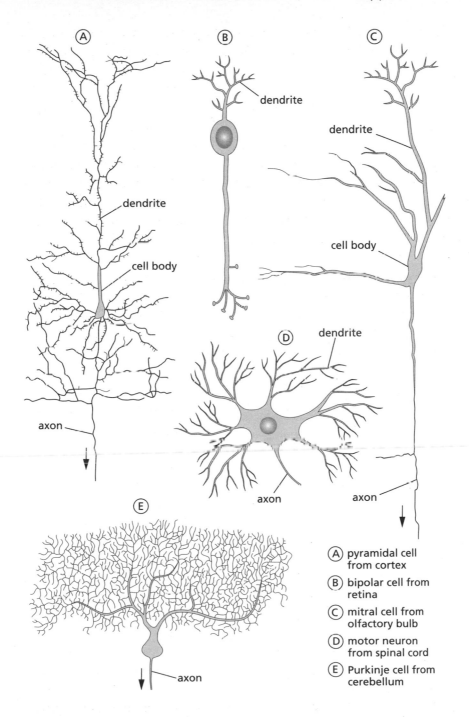

Fig 2.8 *Varieties of neurons.*

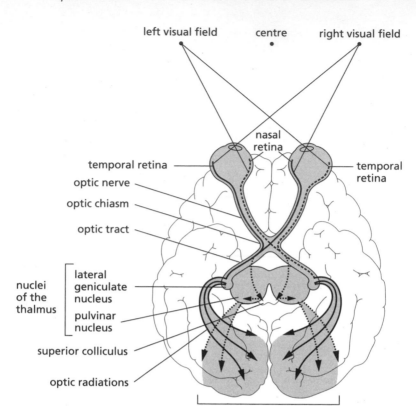

left visual field — centre — right visual field

nasal retina

temporal retina — temporal retina

optic nerve

optic chiasm

optic tract

nuclei of the thalamus — lateral geniculate nucleus — pulvinar nucleus

superior colliculus

optic radiations

visual cortex (occipital lobe)

Fig. 2.9 *The primate visual system – from retina to cortex.*

information arriving from the retina is processed via nuclei in the midbrain (lateral geniculate bodies) from which projections run up to the visual cortex, located towards the back of the cerebrum. However, there is not one single visual area but at least thirty distinct modules, each performing different processing tasks. As well as the cortical layers which receive and transmit information from other brain regions, each column consists of cells capable of detecting specific features of visual information. There are ensembles of neurons that recognise vertical or horizontal lines, others that detect edges, others still that detect colour or motion. There is a similar modular structure to the motor cortex. Feedforward and feedback interactions between these modules can be mapped, and indicate an extraordinarily rich array of interconnections, beside which the simplifications of the London tube map or the wiring on a silicon chip pale into insignificance (Fig. 2.10).

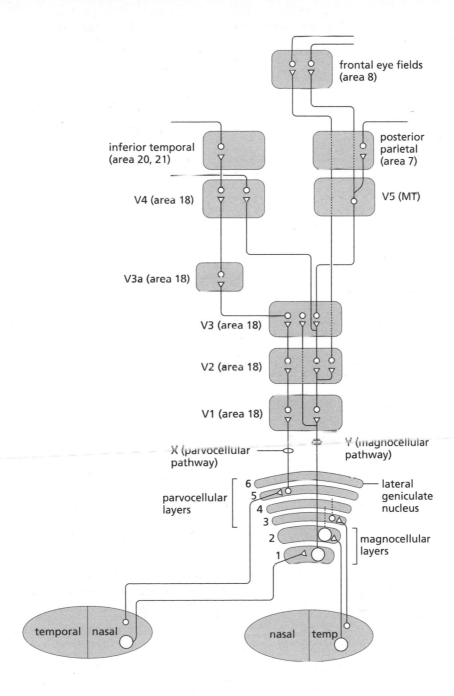

Fig. 2.10 *Visual pathways in the cortex.*

Localised damage to these modules can result in seemingly bizarre anomalies, such as in the case of brain-damaged patients who can detect motion without being able to describe what it is that is moving, or who, having lost their colour analysers, see the world only in black and white.[23] Loss of the entire visual cortex results in seeming blindness, but there seems to be some residual capacity to sense the external world visually without being aware of it – so called blind-sight – presumably reliant on the lateral geniculate, and indicative of the hierarchical way in which evolutionary accretion of 'higher' structures superimposed on early evolving ones leaves some residual independent capacity to the 'lower' regions if cortical control and analysis is unavailable. This is not the time to discuss in any detail the implications of this highly distributed organisation of brain functions, but merely to indicate how it has developed as a result of the continuing tinkering processes of evolution.

Does size matter?

This leads inexorably to an issue that has been the focus of much discussion amongst those concerned with brain evolution. Is there a relationship between brain size and cognitive or affective abilities? Are big brains better? Is there some scale on which humans are top of the brain-tree? What scale would be appropriate? Is there a distinction between actual size and effective size? The attempt to find appropriate scales to relate organisms of different sizes is a general theme in biology, called allometry. Human brains are unusually large by comparison with those of our nearest evolutionary neighbours, the primates (Fig 2.11)

Yet we certainly don't possess the largest brains of any species, weighing in at around 1300–1500 grams (and there is some evidence that early *Homo sapiens* had larger brains than we, their modern successors, do). Elephants and whales have much bigger brains than humans. However, they are also manifestly much bigger overall than we are, and larger bodies by and large should require larger brains because more neurons might be assumed to be required to manage the greater bulk. So maybe brain weight relative to body weight would be a better measure. In general, brain weight and body weight increase in step, but humans are outliers on this curve, with a bigger than expected brain, around 2 per cent of body weight. But even on this score we aren't supreme. The

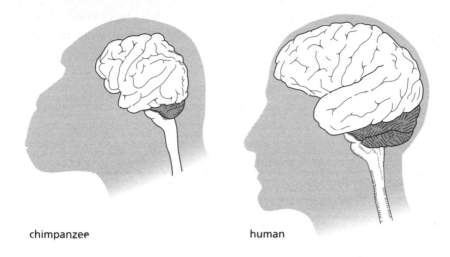

chimpanzee   human

Fig. 2.11 *Chimpanzee and human brains.*

animals I work with in the lab, young chicks, have brains checking in at 4 per cent of body weight, rather like adult mice – twice the brain weight/body weight ratio of humans.

That the bigger-means-better argument is over-simplistic has been neatly demonstrated by the evolutionary anthropologist Terrence Deacon, referring to what he calls 'the Chihuahua paradox'.[24] Chihuahuas are tiny by comparison with, say, Alsatians. Both are of course from the same species, dogs, and in principle interfertile (though the geometry is likely to prove a bit tricky). Both have approximately the same-sized brain, so chihuahuas have a much higher brain weight body weight ratio. Yet they are not regarded as particularly intelligent by comparison with other varieties of dog, so there must be other factors at play.

The mistake that an emphasis on brain weight makes is relatively simple, and derives from the fact that, as I have repeatedly emphasised, the brain is not a unitary organ. Most of the differences among mammalian brains result from the disproportionate growth of later maturing forebrain structures.[25] But amongst the mammals different regions are specialised in respect of specific functions, appropriately adapted to life style. Complex statistical analyses permit a 'cerebrotype' to be defined for each species.[26] Carnivorous mammals for example, have highly developed olfactory capacities, and they need them, for they hunt their prey primarily by smell. By contrast, the primates have a diminished sense of smell but need highly developed eyesight to survive in their arboreal environment, where they must be able to climb, swing and jump, so they have a large cerebellum, concerned with ensuring balance.

Humans, sharing a common ancestry with today's great apes, have a similarly well-developed visual cortex and cerebellum. But also, by comparison with other primates, we possess much enlarged frontal lobes. The basic tool for each of these types of specialisation is the cerebral cortex; which particular areas developed predominantly have depended on the evolutionary niche of the particular organism. However, it does seem to be the case that for each family of mammals, there has been a steady increase in the size and complexity of the cerebral cortex from the first-evolving members of the family to the most recent. Early carnivores had less well-developed cortices than advanced ones, as did early primates compared with more recently evolved ones, including humans. The more recently evolved the organism the greater the volume of the cerebral cortex, and the consequence is that in order to pack in this volume, not only do the cerebral hemispheres increase in size but, as I've said, their area is increased by twisting and convoluting their surface. As a result the number of neurons in the cortex increases.

However, even neuronal number is not a straightforward index. The more complex a system is, the greater the number of higher order control mechanisms that are required. Organisational charts of large companies reveal an inexorable increase in the number and layers of management as the firm's size increases, as junior managers require more seniors (bigger fleas on the backs of lesser fleas, to reverse the old rhyme). True, every so often firms engaged in restructuring exercises claim that they are sweeping out whole layers of management, but somehow they always seem to creep back again (my own university is a powerful case in point). This suggests that such multiple management layers are an inevitable feature of large complex organisations, and we should not be surprised to find that brains (whose evolution does not permit wholesale restructuring anyhow) are no different. In fact I shall be arguing later that the modular and distributed organisation of brain processes minimises rather than maximises the management problem, even though it can't avoid it entirely.

How smart is smart?

How do these changes relate to what might be called intelligence or cognitive ability? Darwin argued that between species differences in intelligence were differences of degree, not of kind. But different species have widely differing skills. Pigeons can be taught to recognise features such as trees or buildings in photographs, to discriminate colours – all skills that are not possible for rats, for instance. Rats will learn to avoid drinking from a water spout if every time they do they get an electric shock to their feet, and they have highly developed olfactory learning skills; chicks can never learn this skill, though they will readily learn to avoid eating food or drinking liquid that tastes bitter. Some bird species – tits and jays for instance – can store many thousands of different food items over the summer and recover them efficiently in the winter when resources are scarce, an ability dependent on their hippocampus. Other birds, zebra finches or canaries for instance, do not have this spatial ability, and don't have such prominent hippocampi, but can learn a large number of subtly different songs, involving the enlargement of several different brain regions. Thus it would appear that amongst vertebrates cognitive skills are strongly associated with life style, or, to put it more formally, ecological niche. However, a more rigorous analysis of these undoubted differences leads to the view that by and large Darwin was right: there are no great qualitative differences in cognition in animal species in the processes of learning and memory.[27]

The whole tenor of this argument is to assume that cognition is all – an easy and conventional assumption to make: that our brains evolved to make us the smartest of all animals, enabling us alone on earth to play chess and even to design computers that can play chess better than we can. But this is not what evolution is about. Above all it is about survival, and strategies for survival – and yes, of course, about differential reproduction so that our offspring survive better in the next generation. Brains, and brainier brains, are about better strategies for survival. As I argued much earlier in this chapter, the evolution of mammals, of predators and prey, involves two things: first, the development of co-operative (as well as competitive) skills within a species, for co-operation can aid both those stalking their prey and those determined to avoid being such prey; and second, a competitive 'arms race' between predators and prey. One aspect of improving both co-operation and competition is to develop bigger and better brains.

Sure, these brains develop cognitive skills, but cognition for a purpose. And what that purpose engages even more than cognition is emotion – referred to more frequently in the trade as affect. Animals learn and remember for a purpose* – to avoid harm (pain, predators, negative affect) or to achieve desirable goals (food, sex, positive affect). This learning is always at the minimum tinged with emotion, and emotion engages more than just the brain – both fear and pleasure involve whole body hormonal responses, including hormones generated in the adrenal glands (steroids such as corticosterone in non-humans, the closely related cortisol in humans) and adrenaline. These hormones interact directly with the brain. Cognition engages the hippocampus, whose neurons are packed with receptors able to interact with steroid hormones, and adrenaline interacts with the brain particularly via another nucleus, the amygdala, triggering neuronal pathways which also activate the hippocampus.

Correcting the assumption that behaving is only about brains, and that cognition is king, is thus important. It also serves as an antidote to a powerful tradition in psychology, and amongst many philosophers,[28] that regards brains as nothing other than sophisticated computers, information processors, cognition machines. As I have hinted throughout this chapter, information is empty without a system to interpret it, to give it meaning. That is why, by contrast with 'information', which implies an absolute measure independent of both signaller and recipient, signalling clearly involves both in mutual interaction. As will become clearer in later chapters, these distinctions are far more than merely semantic, but impact on our entire understanding of brain processes and hence human mind processes (note the appearance of that dreaded word 'mind' for the first time in this chapter!). The cognitive, information-processing obsession goes back a long way in Western philosophy, at least to Descartes, with his famous 'cogito ergo sum' – 'I think, therefore, I am'. This, as Damasio expressed it in the title of his book, is *Descartes' Error*.[29] To understand the evolution of brains and behaviour, and the emergence of humanity, we need at the very least to insist on 'Emotio ergo sum'.[30]

---

* Yes, I know that this statement sounds like that great biological heresy, teleology. It would be possible to restate it to avoid the charge but, frankly, I can't really be troubled with that tedious debate in that context.

The uniqueness of humans?

So, at the end of a chapter tracing the path from the origin of life on earth to the emergence of *Homo sapiens* as a distinct species some 200,000 years ago, what is there that is unique about the human brain which can help us understand the uniqueness of our species? Our biochemistry is virtually identical to that of no-brain species. At even the highest degree of magnification our neurons look the same as those of any other vertebrate; they talk to one another using the same electrical and chemical signals. Our sensory and motor skills are better in some ways, worse in others. We see only in a limited range of wavelengths compared with bees, we hear at only a limited range of frequencies compared with dogs, we are far worse at scenting odours than most carnivores. We can't run as fast as many, or climb as well as others, or swim as well as dolphins. Our brains aren't the biggest, though they do have some unique features such as the relative enlargement of the frontal and prefrontal lobes.

So in what does our uniqueness lie? Partly it is in our versatility. We are the pentathlon experts: we may not be able to do any one of these things as well as others, but we are the only species that can (at least if we are fit enough) run a kilometre, swim a river and then climb a tree. And for sure we are the only species that can then go on to tell others of our kind of our achievements, or write a poem about them. We have above all a deeper range of emotions, enabling us to feel empathy, solidarity, pity, love, so far as we can tell, well beyond the range of any other species (only humans it seems to me, can feel the cross-species empathy that drives many in the animal rights movement – an intriguing paradox!). We have language, consciousness, foresight. We have society, culture, technology. How and why? From where in our evolutionary history do these seemingly unique capacities emerge, and what have our brains to do with enabling them?

These are big questions – too big for this already overlong chapter, but also not answerable from an evolutionary viewpoint alone. I have already expanded Dobzhansky's claim, that nothing in biology makes sense except in the light of evolution, to insist on the need also to encompass development and history. So, in the next chapter, I retrace the argument to look at the emergence of the adult brain, not this time from the point of view of its phylogeny, of evolution, but from the somewhat related one of its embryonic and childhood development, the viewpoint of ontogeny.

CHAPTER 3

From 1 to 100 Billion in
Nine Months

Ontogeny, phylogeny, history

NOTHING, I REPEAT, MAKES SENSE IN BIOLOGY EXCEPT IN THE CONTEXT OF history – and within history I include evolution, development, social, cultural and technological history. The previous chapter has traced how the human brain, a mass of interconnecting cells and pathways, emerged over three billion years of evolution, the product of endless genetic experiments, the inevitable result of the interplay of chance and necessity, of structural and chemical constraints, of the active engagement of organisms with their environment and the selection of those organisms which are best fitted to that environment. Inevitably that account was speculative, derived by using the present as a clue to the past, for we cannot rewind the tape of history. Instead, as I have said, inference is based on evidence from fossils and from the brains and behaviour of those current living forms that seem most closely to resemble them. With development – ontogeny – we are on safer ground. These are processes about which it is possible to speak with some certainty. We can observe human development, and intervene in that of our non-human relatives. The trajectory by which a fusion of human sperm and ovum results, over nine months gestation, in some 3–4 kilos of baby, fully equipped with internal organs, limbs, and a brain with most of its 100 billion neurons in place, is thus relatively easy to describe, even when it is hard to explain.

There's a long rejected idea, due originally to the German Darwinian zoologist Ernst Haeckel in the late nineteenth century, but that one sometimes still hears expressed even today, that ontogeny simply recapitulates phylogeny – the evolutionary path that led to humans.[1] At various stages during its development in the womb, the human foetus is supposed to resemble a fish, an amphibian, an early mammal, in a bizarre retreading of that path. Certainly, there are striking similarities in the appearance of the human foetus to those of other species during its development, although this is less surprising when one recognises that evolution can only work with the materials available to it. Ontogeny must be treated in its own right, not as a sort of evolutionary replay. The concept that we need to keep hold of is that what evolves is not an adult organism but an entire developmental cycle, from conception to maturity and reproduction (what happens post reproductive age is relatively unaffected by evolutionary pressures). However, the overwhelming similarity between humans and non-human mammals, in terms of brain biochemistry, the structure and function of neurons, and indeed the basic architecture of the brain, means that the study of development in non-humans – especially other primates, but even in rats and mice – can tell a great deal about the basic processes involved in our own ontogeny. So whilst the focus of this chapter is on what happens in the foetal human brain during those nine months of gestation, much of the argument will be taken from what can be discovered by observation and experimental manipulation of non-human brains.

Identity and difference

Speaking of the human brain in this universalistic way defines one important truth whilst masking another. All humans are alike in very many respects, all are different in some. (No two individuals, not even monozygotic twins, are entirely identical, even at birth.) Yet chemically, anatomically and physiologically there is astonishingly little obvious variation to be found between brains, even from people from widely different populations. Barring gross developmental damage, the same structures and substances repeat in every human brain, from the chemistry of their neurotransmitters to the wrinkles on the surface of the cerebral cortex.

Humans differ substantially in size and shape, and so do our brains, but when a correction is made for body size, then our brains are closely matched in mass and structure, though men's brains are slightly heavier on average than are women's. So similar are they though, that imagers using PET (positron emission tomography) and MRI (magnetic resonance imaging) have been able to develop algorithms by which they can transform and project the image derived from any individual into a 'standard' brain. Brains are so finely tuned to function, so limited by constraints, that anything more than relatively minor variation is simply lethal.

Within those minor variations, however, must lie also the differences that help constitute the uniqueness of each individual human. That is, our brains demonstrate at the same time the essential unity of humans *and* our essential individuality. The source of both similarities and differences lies in the processes of development, from conception to birth, which take the raw material of genes and environment and employ it in an apparently seamless unrolling. It is easy to assume that this is no more than the reading out of a preformationist program embedded in DNA, relatively unmodifiable – at least until birth by environmental influences. The genetic program, according to this view, instructs development, and then environmental contingencies select which of a number of possible developmental pathways will be chosen. Hence that tired old hangover from nineteenth-century dichotomous thinking, of nature and nurture, genes and environment.[2] Or, in an account once favoured more by psychologists and anthropologists (and some philosophers) than by biologists, biology is in charge until birth, and socialisation and culture thereafter.

Such simplifications are false. 'The environment' is as much a myth as is 'the gene'. Environments exist at multiple levels. Thus for an individual piece of DNA 'the environment' is all the rest of the DNA in the genome, plus the cellular metabolic system that surrounds it, proteins, enzymes, ions, water . . . For a cell in a multicellular organism, as the last chapter spelled out, the environment, constant or not, is the internal milieu in which it is embedded or adjacent cells, signalling molecules, bloodstream and extracellular fluids. For organisms, the environment is constituted by the biological and physical world in which they move – and for humans, the social, cultural and technological world too.

'The environment' impinges from the moment of conception and

(a)

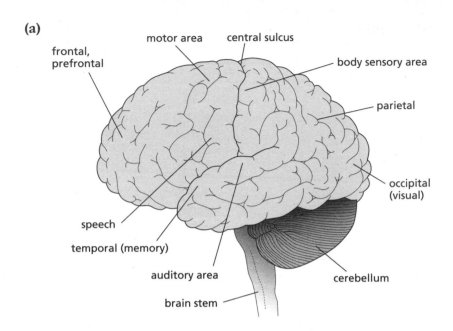

(b)

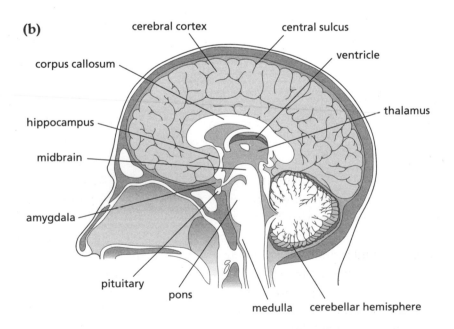

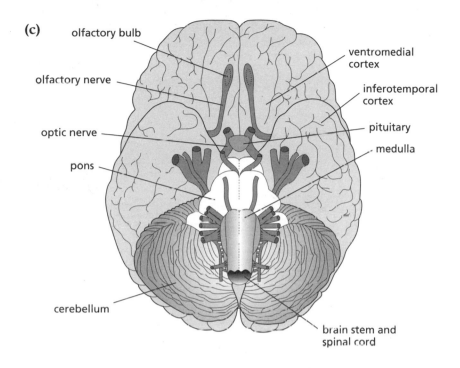

(c) olfactory bulb

ventromedial cortex

olfactory nerve

inferotemporal cortex

optic nerve

pituitary

medulla

pons

cerebellum

brain stem and spinal cord

Fig. 3.1 *The human brain intact (a), midline view (b) and from below (c). The location of some regions and structures described in the text is approximately indicated.*

factors in the maternal womb dependent on the maternal health and context affect development profoundly – even for identical twins, the position of the two foetuses in the womb will ensure developmental differences – but there is more to it than that. The very concept of unpicking genes and environment misspeaks the nature of developmental processes. The developing foetus, and the unique human which it is to become, is always both 100 per cent a product of its DNA and 100 per cent a product of the environment of that DNA – and that includes not just the cellular and maternal environment but the social environment in which the pregnant mother is located. Just as there was no naked replicator at life's origin, there is no genetic 'program' present within the fertilised ovum, to be isolated from the context in which it is expressed. (In an effort to avoid the determinism of the term program, some writers speak of a genetic 'recipe' but I still find this term misleading; the environment, the context in which genes are expressed, is no mechanical cook simply following instructions, but an active partner in development.) The three dimensions of cellular structure and organismic form, and the fourth temporal dimension within which development unrolls, cannot simply be 'read off' from the one-dimensional strand of DNA, like a computer program. Organisms are analogue, not digital.

Autopoiesis

The essential point to grasp is that life is not a static 'thing' but a process. Not only during development but throughout the life span, all living organisms are in a state of dynamic flux which both ensures moment-by-moment stability (homeostasis) and steady change over time, or homeodynamics.[3] There is a paradox about this, beautifully exemplified by thinking about the problem of being a newborn baby. Babies at birth have a suckling reflex – put to the breast, they drink – but within a few months of birth the baby develops teeth and no longer suckles but chews. Chewing is not simply a more mature form of suckling but involves different nerves, muscles and movements. So the problem that all babies have to solve is how, at the same time, to be competent sucklers and to be in the process of becoming competent chewers.

All of life is about *being* and *becoming*; being one thing, and simultaneously transforming oneself into something different. It is really like (just as Dawkins and others have commented, in the context of evolution) rebuilding an aeroplane in mid-flight. And we all do it, throughout our lives – not just babies but adults, not just humans but mice and fruitflies and oak trees and mushrooms. That is why I argue that living creatures are continually constructing themselves. This process is one of self-creation, autopoiesis,[4] or (as it has sometimes been called) developmental systems theory.[*5] The cell, the embryo, the foetus, in a profound

---

* My friend and colleague Kostya Anokhin has pointed out that the concept of autopoiesis reprises the earlier term, systemogenesis, introduced in the 1930s by Soviet neuropsychologists (including his grandfather Peter Kuzmich Anokhin) and developmental biologists working within a well articulated 'dialectical' Marxist theoretical framework. As I've pointed out previously, modern neuroscience – and most specifically Anglo-Saxon pragmatic reductionism – ignores these earlier insights to its disadvantage.

sense 'chooses' which genes to switch on at any moment during its development; it is, from the moment of fertilisation, but increasingly through that long trajectory to birth and beyond, an active player in its own destiny. It is through autopoiesis that the to-be-born human constructs herself.

Of no body organ is the developmental sequence more simultaneously dramatic and enigmatic than the brain. How to explain the complexity and apparent precision with which individual neurons are born, migrate to their appropriate final sites, and make the connections which ensure that the newborn on its arrival into the outside world has a nervous system so fully organised that the baby can already see, hear, feel, voice her needs, and move her limbs? The fact that this is possible implies that the baby at birth must have most of her complement of neurons already in place – if not the entire 100 billion, then getting on for that number. If we assume a steady birth of cells over the whole nine months – although of course in reality growth is much more uneven, with periodic growth spurts and lags – it would mean some 250,000 nerve cells being born every minute of every day over the period. As if this figure is not extraordinary enough, such is the density of connections between these neurons that we must imagine up to 30,000 synapses a second being made over the period for every square centimetre of newborn cortical surface. And to this rapid rate of production must be added that of the glia, packing the white matter below the cortex and surrounding the neurons within it – though admittedly they do not reach their full complement by birth but continue to be generated throughout life.

Two questions follow, which it is the purpose of this chapter to answer insofar as current knowledge permits. First, how does the dynamic of development account for the seeming invariance of the human brain, progressing from egg to embryo to foetus to child with such extraordinary precision? Second, how can it explain the differences between brains which have so developed? Both are immanent within the process of autopoiesis. The first of these, invariant development within a fluctuating environment, is termed specificity; the second, the variations that develop as adaptations to environmental contingencies, is plasticity. Much of what needs to be understood about the brain is embraced within these two intertwined processes, a developmental double helix, if you like. Without specificity, the brain would not be able to become accurately wired, so that, for instance, nerves would not be able to make the right connections en route from retina to visual cortex to enable binocular vision, or from motor cortex via the spinal cord to innervate muscles. But without plasticity, the developing nervous system would be unable to repair itself following damage, or to mould its responses to changing aspects of the outside world so as to be able to create within the brain a model, a representation of that world and a plan of how to act upon it which it is the function of brains to provide. It is specificity and plasticity rather than nature and nurture that provide the dialectic within which development occurs, and both are utterly dependent on both genes and environment.

Whilst I am in the business of dismantling dichotomies, there is yet one other that needs abandoning before one can go much further, and that is the 'brain/body' distinction. The previous chapter described how neuronal connectivity evolved from earlier hormonal signalling systems, and how diffuse ganglia – ensembles of nerve cells – became concentrated at the head end to form brains. But the intimate interconnectivity of brains and bodies remains. Indeed there is an entire, though often unregarded, nervous system in the gut – the enteric nervous system – with getting on for as many neurons as the brain itself (yes, we do sometimes think, or at least feel, with our bowels, as the ancient Hebrews maintained). Brains with their voracious demand for glucose and oxygen are utterly at the mercy of the body's circulatory system to provide them. Nervous processes originating in the lower regions of the brain, but generally not requiring cortical supervision, can regulate heart rate and breathing. These intimate mutual regulatory processes occur at many different levels. The hypothalamus regulates the release of hormones from the pituitary, hormones amongst whose major functions is the regulation of the release of *other* hormones from the adrenal glands, testes and ovaries. But reciprocally, the neurons in many brain regions (including hippocampus, hypothalamus and amygdala) carry receptors that respond to steroid hormones such as cortisol, peptide hormones such as oxytocin and vasopressin, and indeed adrenaline.

Whilst such interactions are relatively well understood, there are many others that are only just beginning to come to neuroscience's attention – notably the complex interplay between brain and immune system, which is giving rise to an entire new research field of psychoneuro-immunology. No account of the development of the brain can be complete without understanding that what we need to interpret is not just the growth of an organ but of an organism of which the organ in question – important though the brain is – cannot be understood in isolation. No brains in bottles here.

Building brains

The problem is not simply that of an enormously rapid production of cells. The brain is a highly ordered structure; neurons have to know their place, to recognise to whom they are supposed to be talking, ensuring that dendrites and axons make the proper connections. To appreciate the problem, remember that the brain is far from homogeneous. It is divided into numerous functionally specialised regions – mini-organs – and in each region the cells are laid down in a strictly controlled pattern. Wiring this structure up correctly would be a tough enough problem if each neuron at its birth was already in the appropriate location – embryonic cortex or thalamus or wherever – but this is not the case. The cells are born in one location – a sort of neuronal maternity ward – and as they mature they leave home in search of destiny, migrating accurately over huge distances.

The developmental history of the brain starts from the moment of conception. The fertilised ovum begins to divide within about an hour; one cell becomes two, two four, four eight . . . Within about eight hours the dividing cells have formed a hollow ball one cell thick and containing about a thousand cells in all (or ten rounds of cell division). Then the cell-ball begins to change shape, as if it were being pushed in at one point until the indentation reaches the inside of the opposite wall of cells, resulting in two hollow balls of cells – the gastrula (Fig. 3.2). As division proceeds the gastrula twists and turns, develops further indentations, regions become pinched off entirely to form independent structures, and within surprisingly few further cell divisions – after all, it only takes twenty divisions for a single cell to multiply to over a million – the cell mass is recognisable as a miniature version of the adult.

The central nervous system originates as a flat sheet of cells on the upper (dorsal) surface of the developing embryo, the so-called neural plate, which, by the time the embryo is some 1.5mm in length, folds in on itself to become the neural groove. It was the appearance of the

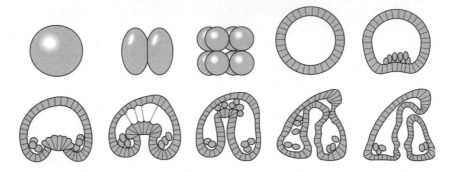

Fig. 3.2 *Cell division, from the single fertilised ovum to the multicellular gastrula.*

groove at this time in development that led to the compromise adopted by the UK's Human Embryology and Fertilisation Committee's report[6] in 1985 which allowed embryos to be grown and manipulated *in vitro* for fourteen days – the argument being that the first appearance of a recognisably neural structure represented the emergence of potential individuality and sentience. The evidence for such an assertion is, to say the least, questionable, but it did allow for a bioethicist-approved compromise between the wishes of the researchers and the concerns of the more theologically minded on the committee.

In any event, in the days following the appearance of the groove the forward end thickens and enlarges – it will later develop into the brain – while, also at the head end of the embryo, other surface regions swell and begin the path of differentiation which will lead to eyes, ears, and nose. Even at 1.5mm in length, the embryo is gathering itself together for its spurt towards sentience. As development proceeds, the neural groove deepens: its walls rise higher. Then they move towards one another, touch, and seal over: the channel has become a tunnel. The neural groove is now the neural tube, several cells thick. By twenty-five days, when the embryo is about 5mm long, the process is complete. The tube is buried along its entire length, and begins to sink beneath the surface of the embryo until it is embedded deep inside. The central cavity will become, in due course, the central canal of the spinal cord, which, at the head end, expands into the ventricular system: fluid-filled spaces in the interior of the brain, their inner surfaces bathed with a continual soothing flow of cerebrospinal fluid (Fig. 3.3).

As cell proliferation continues, the foetal brain continues to grow in

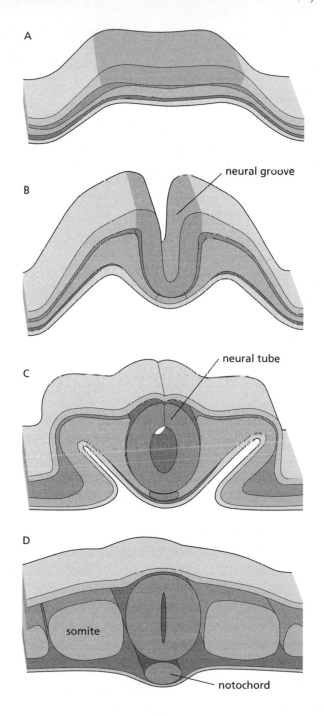

Fig. 3.3 *Trajectory from neural groove to brain.*

size. Three swellings appear in the neural tube, due in time to become the forebrain, midbrain and hindbrain (this is one of the features that persuaded the early embryologists into the 'ontogeny recapitulates phylogeny' belief). By the time the embryo is 13mm in length, this early three-vesicle brain has become a five-vesicle brain, with the forebrain differentiating to separate the thalamic region (diencephalon) from the telencephalon, the region eventually to become the cerebral hemispheres, which are already apparent as two bulges on either side of the tube (Fig. 3.4). These structures are already becoming too large to accommodate as part of a straight tube, and the tube bends back on itself with two distinct kinks, one towards the base, the other at the midbrain region. Although these curves change in orientation as development proceeds, they are already beginning to locate the brain in a characteristic position with respect to the spinal cord, with a ninety-degree bend in its axis. Beyond the hindbrain, the neural tube thickens to become the cerebellum. Meanwhile, at the level of the forebrain, two little outgrowths appear, one on either side, which grow towards the surface into goblet-like shapes connected by stalks to the rest of the brain. They are the optic cups, which in due course develop retinae. The stalks form

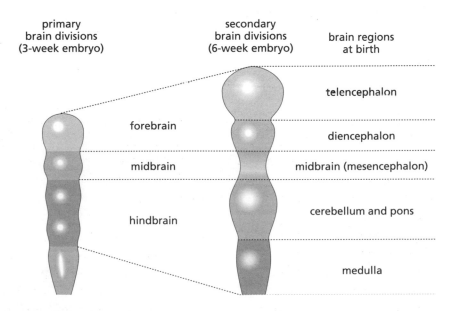

Fig. 3.4 *The five-vesicle brain.*

the optic nerves, and the eyes, developmentally parts of the brain, have emerged.

From now on the appearance of recognisably brain-like features is rapid. By the end of the third foetal month, the cerebral and cerebellar hemispheres can be plainly traced, while the thalamus, hypothalamus and other vital centres can also be differentiated. In the following months the cerebral hemispheres swell and extend. By the fifth month the cortex has begun to show its characteristic wrinkled appearance. Most of the main features of the convolutions are apparent by the eighth month, although the frontal and temporal lobes are still small by comparison to the adult, and the total surface area of the cortex is much below its eventual size.

Below the brain the neural tube becomes the spinal cord. The links between the spinal cord and the emergent musculature – the peripheral nervous system – develop early, often before the muscle cells have themselves migrated to their final position. The muscles arise fairly closely adjacent to the neural tube, and, once the nerve connections are formed, they move outwards, pulling their nerves behind them like divers their oxygen lines. Once some nerves have formed primary links in this way, others may join them as well, following the path traced by the original nerve and finally spreading to adjacent tissue.

The birth of cells

Over these few months of embryonic and foetal development all the billions of neurons and glia which must ultimately constitute the brain begin to spin off from the neural tube. It is the cells close to the surface of the ventricles which are destined to give rise to the neurons and some of the glial cells. In the first stage the cells of the neural tube become specialised into two broad types, precursors of neurons and glia respectively – neuroblasts and glioblasts. Cells in this stage of differentiation can be extracted from the embryos of experimental animals – chick embryo is a favourite source – and maintained alive, growing and dividing *in vitro* through many thousands of generations, for the neuronal precursors, the neuroblasts, have not yet lost their capacity for division. This comes later in their development when they turn into full-grown neurons. The baby neurons can be put into a saucer full of nutrient

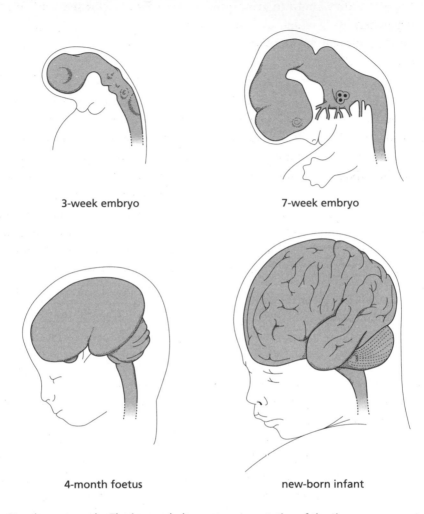

Not drawn to scale. The images below are representative of the sizes in relation to the drawn size of the new-born infant's brain

Fig 3.5 *Development of the human brain.*

jelly, where they begin to grow, put out axons and dendrites and search for partners with which to make synapses.

Information about the birth and migration of such newborn cells was extremely difficult to obtain before the advent, in the 1960s, of the technique of autoradiography. Because each new cell requires its complement of genes, as each cell is born strands of new DNA must be synthesised from precursor molecules, including the thymidine (which provides DNA's T). If radioactively-labelled thymidine is injected into a pregnant mammal (routinely a rat or mouse), it becomes incorporated into the DNA in the foetal cell's nucleus and, as the cells are born and begin to migrate, the radioactive DNA becomes distributed through each of the daughter cells. If the brain is then cut into thin sections and placed next to a sheet of photographic film, the radioactive sites will blacken the film they touch, thus creating a map of the distribution of the radioactivity. Depending on when during foetal development the radioactive thymidine is injected, the birthdays and subsequent fate of individual cells and their progeny can be mapped.

During the first hundred days of human foetal life the majority of neurons will be born in the ventricular zone. The vast majority of glia are produced by cells in a proliferative layer further from the ventricles (the subventricular zone), which gradually takes over the generation of cortical cells. The newborn neurons migrate from the ventricular zone towards what will become the cerebral surface. There, the young neurons meet axons, growing in from regions of the developing midbrain, and establish the first horizontal layering. The later-born neurons have to pass through this region, eventually to form layers 2 to 6 of the adult cortex.[7] They thus assemble in an 'inside-out' order – that is, the deepest cellular layers are assembled first and those closest to the surface last.

The implications of this sequence are that the young neurons are required to migrate from this site of origin to their ultimate locations, distances which may be tens of thousands of times their own length – equivalent to a human navigating over a distance of twenty-five kilometres. How do they find their way? Does each cell know where it is going and what it is to become before it arrives at its final destination – that is, is it equipped with a route map – or is it a general purpose cell which can take on any appropriate form or function depending on its final address within the brain? The debate over whether the newborn cell carries instructions as to its ultimate destiny, or whether its fate is determined – selected – by the environment in which it ends up, has been fierce. It is of course a version of the nature/nurture argument; as such it is a similar false dichotomy, resulting in part from the fact that the great expansion of genetic knowledge of the last decades has not yet been matched by a comparable increase in the understanding of development.

Patterns of migration

Young neurons must recognise their migratory path and move along it; finally they must at some point recognise when to stop migrating and instead begin to aggregate with other neurons of the same kind, put out axons and dendrites, and make the right synaptic connections. It turns out that glia have a vital role to play in helping the neurons along their way. There are specialised glial cells within the developing cortex whose cell bodies are close to the ventricles and whose fibres extend radially to the surface of the cortex. These cells, called, unsurprisingly, 'radial glia', form an advance party for the neurons, moving away from their sites of origin and towards what will become the cortex and, as they do so, creating part of the scaffolding along which the neurons can move.

As the glia migrate they spin out long tails up which the neurons can in due course climb. The cell membranes of both neurons and glia contain a particular class of proteins called CAMs (cell adhesion molecules). In the developing tissue the CAM work a bit like crampons; they stick out from the surface of the membrane and cling to the matching CAM on a nearby cell; thus the neurons can clutch the glia and ratchet themselves along. As a further trick the migrating cells also lay down a sort of slime trail of molecules related to the CAMs – substrate adhesion molecules or SAMs – which provide additional guidance for the cells following (Fig. 3.6). The neurons can move along this SAM-trail rather like amoeba – that is, they ooze. At the forward end they develop a bulge – a growth cone, from which little finger-like projections (filopodia) are pushed out to probe for the molecules they can recognise, to provide the guidance they need. In neurons maintained alive in tissue culture, time-lapse photography shows the filopodia tentatively extending and retracting, using an internal musculature of actin, as they sniff their surroundings for map-references to direct their growth.

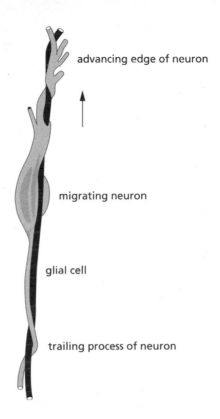

advancing edge of neuron

migrating neuron

glial cell

trailing process of neuron

Fig. 3.6 *A migrating neuron with a glial cell as guide.*

What provides such map-references for the cellular route-marches? Even if neurons follow glia, the glia themselves must have a sense of direction. Both distant and local signals must be involved. One way of signalling direction is to have already in place some target cell or tissue towards which the migration must be directed. Suppose the target is constantly secreting a signalling molecule, which then diffuses away from it. This will create a concentration gradient, highest at the target and progressively weaker at increasing distances from it, just as in the case of the unicells discussed in the last chapter. In the 1950s Rita Levi-Montalcini identified one such signalling (or trophic) molecule, which she called nerve growth factor (NGF); by the time she was awarded her Nobel prize for the discovery, in 1986,[8] it was recognised as but one of a growing family of such molecules. Whilst NGF is important in the peripheral nervous system, at least two others, brain-derived neurotrophic factor (BDNF) and glial-derived neurotrophic factor (GDNF), play the crucial guidance roles within the brain itself. In addition, early-born neurons secrete a protein called reelin which sticks to the molecular matrix surrounding them, acting as a stop signal for each wave of arriving cortical neurons, telling them to get off the glial fibre and develop into a layer of mature neurons.

Trophic factors can provide the long-range guidance by which the growing axons of motor nerves can reach out and find their target muscles, or the axons from the retinal neurons which form the optic nerve can track their way to their first staging post within the brain, the lateral geniculate. However, the migrating cells or growing axons also need to keep in step with one another; each has to know who its neighbours are. The diffusion of a local gradient molecule, together with the presence of some type of chemosensors on the axon surface, could enable each to determine whether it has neighbours to its right and left and to maintain step with them.[9] The entire troop of axons would then arrive in formation at the lateral geniculate and make appropriate synaptic connections, thus creating in the geniculate a map, albeit a topographically transformed one, of the retina – not exactly a camera image, but its equivalent. Indeed the brain holds many such maps, multiple maps for each of its input sensory systems and output motor systems, maps whose topology must be preserved during development (Fig. 3.7).[10]

Instruction and selection

The process just described would be compatible with a model in which each axon is kept on course by instructions from its environment, both the trophic factor diffusing from the target region and the relationships with its nearest neighbours directing it to its final site. There is indeed some evidence that a considerable part of nervous system development can be contained within such a model.[11] However, there is a further process operating here. During embryonic development there is a vast overproduction of cells. Many more neurons are born than subsequently survive. More axons arrive at their destination than there are target cells to receive them. They must therefore compete for targets. Those that do not find them wither away and die. There is thus in this model of development, competition for scarce resource – trophic factor, target cell, synaptic space.

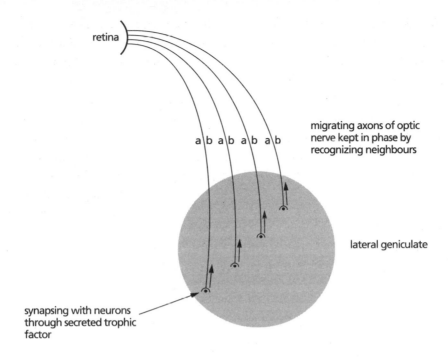

retina

a\b a\b a\b a\b

migrating axons of optic
nerve kept in phase by
recognizing neighbours

lateral geniculate

synapsing with neurons
through secreted trophic
factor

Fig. 3.7 *Guiding factors in axonal growth.*

Where axons meet either the dendrite or cell body of another neuron, synapses can begin to be built. Once again this process can be studied by time-lapse photography in culture, or even, with some ingenious new techniques, by real-time imaging of living neurons *in situ*. It appears that the various components required to build synapses are synthesised in the cell body and transported, pre-packaged, so to say, down the axon to accumulate at potential presynaptic sites within one to two hours of initial contact between the cells. The presence of these pre-synaptic components, and release of neurotransmitter molecules, helps trigger a complementary process on the post-synaptic side, as the specialised receptor molecules for the neurotransmitters begin to be inserted into the dendritic membranes, transforming dendritic filopodia into the spine-like protrusions which stud the surface of mature dendrites and ensure efficient contact between pre- and post-synaptic cells.[12] This process is highly dynamic; synapses are rapidly formed, but can also as rapidly be eliminated and their constituent molecules recycled in the absence of active interaction between pre- and post-synaptic neurons.

During development there is thus a superabundance of synaptic production, a veritable efflorescence – but if synapses cannot make their appropriate functional connections with the dendrites of the neurons they approach, they become pruned away and disappear.

This overproduction of neurons and synapses might seem wasteful. It has led to the argument that just as during evolution 'natural selection' will eliminate less-fit organisms, so some similar process of selection occurs within the developing brain – a process that the immunologist and theorist of human consciousness Gerald Edelman has called 'neural Darwinism'.[13] However, this transference of the 'survival of the fittest' metaphor from organisms to cells is only partially correct. It seems probable that the whole process of cellular migration over large distances, the creation of long-range order, requires the working out of some internal programmes of both individual cells and the collectivity of cells acting in concert. Even though synapses from only a particular neuron may end up making successful connections with its target cell, if the others had not been present during the long period of growth and migration it is doubtful whether a single nerve axon would have been able even to reach the target. Survival of one depends on the presence of the many. Overproduction and subsequent pruning of neurons and synapses may at one level of magnification look like competition and selection; viewed on the larger scale, they appear as co-operative processes. It thus seems to be necessary, to ensure that enough cells arrive at their destination and make the relevant connections, that others must assist them, only themselves to die en route – a process called programmed cell death or apoptosis. Indeed, in a creature like *C. elegans*, in which the developmental process can be mapped out in exquisite detail and the function of each neuron is known, it is clear that apoptosis is not a matter of simple competition and selection. Those neurons destined to be born and to die en route to their seemingly final destination can be identified from the start. The process is orderly rather than random. Apoptosis is a fundamental and necessary part of development.

It used to be thought that, by contrast with the glia, the neuronal population of the human brain was virtually complete at birth, bar a few additions to regions of the cortex such as the frontal lobe, and that after birth it is downhill all the way. As neurons die, it was assumed, the spaces they occupy become filled by glia and the branching dendrites of still-surviving neurons. However, it is now clear that this isn't entirely true. Although there is a loss of neurons with age, especially in later years, even as an adult the brain (and especially the olfactory cortex) holds in reserve a small population of so-called progenitor, or stem cells, with the capacity to differentiate into neurons when required. This potential is currently the focus of intense research as it may offer the prospect of repairing brains damaged by disease or injury – a theme that I will return to in later chapters.

Genealogy, fate and autopoiesis

Cells in the ventricular zone face many choices in early development. To start with they are said to be totipotent – capable of all possible futures. As they mature they slowly lose this capacity, first becoming pluripotent – perhaps with their future as a neuron fixed, but as what sort of neuron still indeterminate. Then, as their future becomes increasingly restricted, the decision remains to be made as to which type of neurotransmitter either to produce or to respond to of the many available. There was once a famous dictum which claimed one neuron/one-neurotransmitter. However like many such 'laws' in biology it has proved to be an over-simplification. The cellular DNA allows for all of these potentials; the ultimate fate of any newborn cell is, however, not for the DNA to determine.

But at what point, and why, during its maturation does a cell become committed to turning into a neuron rather than a glial cell? If it decides to become a neuron, when does it decide to become a pyramidal or a basket cell? Will the neuroblast continue to proliferate after this decision is made? Is a particular cell programmed to reside in a certain layer before it makes its migratory journey? And at what stage of development is the cell's potential constrained to forming a particular set of connections and producing a certain neurotransmitter? Knowing the stage at which a cell becomes committed to adopting a certain final form enables one to begin to narrow down the factors that might be involved in this developmental process. However, interpreting the time-course and mechanism whereby a neuron becomes thus committed has proved hard going, because of both the brain's intricate structure and its inaccessibility during embryonic development. Even so, enough is known to make it clear that cortical neurons acquire their final form through the classical interplay of genes and environment within autopoietic development.

Clues as to the processes involved come from a study of cell lineages – that is, the routes by which progenitor cells eventually become mature glia and neurons via a sequence of cell divisions. An instructional model would argue that whether a cell became a neuron – and if so whether it was a pyramidal or a stellate neuron – was pre-ordained at its birth, specified in its genetic programme. But of course, all cells in the body contain identical DNA, and how a cell matures will depend on which particular regions of DNA – which genes, therefore – are switched on and off at any particular time. It is here that the context of development becomes important, by regulating gene expression during the migratory and maturational steps. So does this mean that the environment selects the cell's fate – the apparent alternative to the instructional model?

As I have already insisted, this way of dichotomous thinking is mistaken; the interplay of development transcends both instruction and selection, genes and environment within the autopoietic process. Manipulating genes in the developing rat embryo, for instance, shows that both the address at which you are born and your family history – your genealogy – is important. Developmental biologists can construct what are portentously called 'fate maps' by which the history of developing cells can be tracked back to their birth and parenthood. Such methods show that clusters of adjacent newborn cells (called clones – one of the many meanings of that overused word) in the ventricular zone are already dedicated to become either glia or pyramidal or non-pyramidal neurons, and to migrate to approximately the same region of the cortex. However, as they migrate they interact with a multitude of external signals: neurotrophic factors, neurotransmitters, neuropeptides and extracellular matrix molecules derived from many sources, axons from elsewhere in the developing brain, or the cerebrospinal fluid, to name but a few possibilities. The immature cells have receptors on their surface capable of recognising and responding to such factors.

The search for such signals and their precise role in the development of the cortex has intensified in recent years. For instance, noradrenaline, dopamine and serotonin are present in the developing brain long before they can be expected to function as neurotransmitters, and one at least, serotonin, has been found to promote the survival and differentiation of cortical pyramidal neurons. Another growth factor (bFGF) helps keep cortical progenitor cells in a proliferative mode, but it has no effect on the rate at which the cells divide. In addition, the differentiation of neuronal, but not glial, progenitors is delayed by the presence of bFGF, thus confirming that proliferation and neuronal differentiation occur sequentially during development. Interestingly, bFGF also stimulates the expression of receptors for a particular inhibitory neurotransmitter, gamma-amino-butyric acid (GABA) in progenitor cells in the ventricular zone of the developing cortex. (Inhibitory neurotransmitters are those that act so as to diminish the chance of the postsynaptic neuron firing.) GABA, produced in already differentiated cells, provides a feedback signal that terminates further cell division and promotes cell differentiation, perhaps by stimulating the expression of particular genes which in turn induce neuronal differentiation. It thus forms part of the regulatory mechanism by which specific subclasses of neurons with their characteristic shapes and patterns of neurotransmitter expression are generated.

Developing function

Up till now the focus of this chapter has been on the development of brain structures, but what really matters is how this ontogenetic process relates to the development of behavioural competence, of the transformation of a fertilised ovum into a competent human individual. Starting from the assumption that performance must be supported by an appropriate brain structure, how far can one analyse the development of function, of behaviour, against the emergent structure of the brain?

The birth and migration of neurons and glia, the development of a five-vesicle brain, cortex and connections of nerves to muscles, and sense organs to brain, is part of preparation for the drama of birth and the increasing autonomy of the human infant. However, there is more involved here than simply wiring up the structures and then pressing the On-switch at the moment of birth. Function develops in parallel with structure during the nine months of human pregnancy, and indeed there is an important sense in which function determines structure – for instance in the way that synapses retract and die if they are not activated by signals running between the neurons.

Some functions appear very early in development. Indeed the blood circulation system and the nervous system are the first to function in embryonic life, with heartbeat commencing in the third week following conception. One of the key measures of the life of the brain, of its functional maturation, is the development of electrical activity, that index of trans-neuronal trafficking and communication. The sum of all this traffic can be recorded as an electroencephalogram (EEG), measured in adults by putting a mesh of recording electrodes on the head and analysing the global signals. Making such recordings prior to birth is of course a rather more complex procedure, but can be achieved non-invasively by placing recording electrodes on the pregnant mother's belly. EEGs can also be recorded in prematurely born babies maintained in incubators – a process designed to monitor their development.

As far back as the third pre-natal month very slow waves of electrical activity can be detected at the surface of the brain, and continuous electrical activity in the brain stem seems to occur from the fourth month onward. These electrical measures are signs that neurons are signalling to one another, that at least some synapses are functioning, and patterns of brain activity are becoming co-ordinated in space and time long before birth. The earliest patterns that have been recorded in this way have been from 24- to 28-week-old prematurely born babies, and it is found that a striking change takes place over this period of time. Before twenty-eight weeks the patterns are very simple and lacking in any of the characteristic forms which go to make up the adult pattern. At twenty-eight weeks the pattern changes, and bursts of regular waves occur, closer to the adult pattern in which characteristic frequencies (called alpha, theta etc. waves) can be detected. All this activity is still only periodic, occurring in brief spasmodic bursts, but after thirty-two weeks the pattern becomes more continuous, and characteristic differences begin to appear in the EEG pattern of the waking and sleeping infant. By the time of normal birth the pattern is well developed, although certain characteristic and specific differences between children appear, partly depending upon the condition of the individual child at birth. From then on the EEG pattern gradually approaches that of the adult, but doesn't reach full maturity until around puberty.

Such patterns of electrical activity are matched by the development of simple reflex arcs with appropriate synaptic connections and interneuronal activity being brought into play in order to co-ordinate muscular movements. The area of sensitivity which will evoke reflex activity gradually expands. At eleven weeks the foetus will make swallowing movements if the lip regions are touched, at twenty-two weeks it will stretch out its lips and purse them, and at twenty-nine weeks it will make sucking movements and sounds. By twelve weeks it will kick and by thirteen the diaphragm muscles will produce the breathing sequence. The whole gamut of simple and more complex reflex systems is brought into operation as birth nears.

Developing difference: sex/gender

Despite the opening paragraphs of this chapter which talked of both similarities and differences, the story so far has been a universalistic one, of generic 'normal' brain development. But brains develop differently from one another even in the womb. Each is unique, genetically and environmentally, even prior to birth as the foetus begins the task of self-construction. The general pattern of neuronal wiring, of synaptic connections, of cortical sulci and gyri, of modular neuronal columns is universal, but the specificities are individual, the constructs of every about-to-be-human foetus's developing lifeline. The complexities of interpreting such differences are of course the stuff of controversy, none greater than that surrounding sex and gender, insofar as these can be disentangled. After all, 'boy or girl?' is likely to be the *second* question asked by the parents of any newborn baby, although the answer isn't necessarily as straightforward as it sometimes seems.

Sex (as opposed to gender) begins at conception. Of the twenty-three pairs of chromosomes containing the DNA we inherit, one pair differs from the start. In normal development, females have a pair of X chromosomes, males one X and one Y. So the patterns of genetic inheritance also vary between the sexes from the outset. (There are also some abnormal patterns of inheritance such as Turner's Syndrome, where a female inherits only one X chromosome, and XYY, in which males inherit an extra Y.) What differences do such normal or abnormal patterns make to brain development? As I've said, on average, males tend to have slightly heavier brains than females. On the other hand, female babies tend to be 'more advanced' in terms of behaviour and functional capacity at birth than do males. During post-natal development other tiny differences in brain structure appear. Although the two cerebral hemispheres appear identical in size and shape, there are small asymmetries, and male brains tend to be more asymmetric than female. There are also claimed to be differences in structure and size in regions of the hypothalamus and corpus callosum. This is, not surprisingly, a fiercely disputed area, because of the ways in which both similarities and differences can be appropriated (or misappropriated) to claim explanatory power in explaining the persistence of the domination of men in a patriarchal society and I will return to it in later chapters.[14]

It is well known that certain traits, such as colour-blindness and haemophilia, can be carried through the female but are only expressed in the male, a consequence of the XX/XY difference between the sexes. The commonplace statement is that all human brains begin as female, and that some 'become masculinised' during foetal development, a 'masculinisation' for which a chromosomal difference is necessary but not sufficient. A key to this 'masculinisation' is the hormone testosterone. Popularly, testosterone is the 'male' hormone, oestrogen the female. However, in reality testosterone and oestrogen are produced and responded to by both males and females; it is only the proportions that differ, testosterone being on average present in higher concentration in males. Neither hormone is actually made in the brain, but both can enter through the blood stream, and there are receptors on neuronal membranes in the hypothalamus and other brain regions that recognise the hormones. What 'masculinises' the otherwise female brain is a surge of testosterone production occurring between eight and twenty-four weeks into pregnancy. This is part of the process of differentiation between male and female brains, with characteristic average differences appearing in the distribution of neuronal receptors for the hormones.

The existence of receptors in the brain for testosterone and oestrogen produced in other body regions also illustrates the importance of the multiple brain–body interactions mentioned earlier on in this chapter. The sex hormones are not the only steroids that affect brain processes and they are also closely chemically related to the brain's own equivalent of steroid hormones, the neurosteroids, which act a bit like BDNF and other growth factors, but which are also present prior to birth in different concentrations in male and female brains. These complex hormonal interactions, occurring even pre-natally, are but one of many reasons why it isn't possible simply to 'read off' average differences between boys and girls, men and women as 'caused by' genetic and chromosomal sex differences, and why the relationship between understanding sex and understanding gender has been so tricky. Such differences are indeed average, and there is considerable variation, which is just part of the problem of attempting to reduce human differences in sex and gender (and still more in sexual orientation) to simple statements about chromosomes, hormones, or any other unilinear 'biological' measure.[15]

Developing individual differences

There's another odd feature of the story of brain development in relation to sex which has recently come to fascinate neuroscientists, this time relating to the sex not of the foetus but of the parent. It was traditionally assumed that because we all inherit two variants of each gene (alleles), one from each parent, it didn't make any difference which parent supplied which allele – but it turns out that this isn't quite true. There is a phenomenon called genomic imprinting, which ensures that in some brain regions it is the allele inherited from the father, and in others the allele from the mother, which is expressed. It seems that in cortical regions the maternal, and in midbrain regions the paternal, alleles are the active ones, the corresponding allele from the other parent being silenced.[16] The implications of this odd phenomenon have provided a field day for evolutionary speculation, but as such unprovable just-so stories can be multiplied indefinitely according to taste, they are scarcely worth considering until the phenomenon is rather better understood.

More important is to understand the ways in which the unrolling of the developmental programme contributes even pre-natally to individual differences. It begins, of course, with the genetic uniqueness of each individual. Most of such genetic differences that are understood are, for obvious reasons, those that result in serious abnormalities in development of brain and behaviour – sometimes resulting in miscarriage or early death, sometimes as neurological diseases that manifest themselves only later in life, such as Huntington's Disease, which is the consequence of a single gene defect (though one that can take a variety of forms) or Alzheimer's disease, where particular combinations of genes represent risk factors that can increase or decrease the likelihood of developing the disease in old age.

Better understood, though also mainly for their negative effects, are aspects of the health and life of the pregnant mother that can affect brain development in the foetus. These include infections and obvious nutritional factors, as the foetus is entirely dependent on a placental supply. Thus severe malnutrition, particularly in the earliest stages of pregnancy around conception, can result in lower brain weights and cognitive deficits in the baby (and of course an increase in miscarriages). However, if the malnutrition occurs later in pregnancy some catch-up is possible. Deficiency of the vitamin folic acid, especially early in pregnancy, can result in neural-tube defects, in which proper closure of the tube fails to occur. If this is at the lower end of the tube, the spinal cord fails to become properly enclosed within the spine, resulting in spina bifida; if it is at the head end, the brain does not develop properly (anencephaly). The mother's use of alcohol, cigarettes or other centrally acting drugs, legal or illegal, can also contribute to cognitive impairment in the baby. On the other hand, severe maternal stress can have similar effects. Stress alters hormone balance, notably of steroid hormones, and elevated levels of, for instance, cortisol, will cross the placenta, interact with receptors in the foetal brain and lastingly change the pattern of development. Such stress can result from economic and social insecurity and lack of stable emotional support – all-too-frequent conditions in our unstable and poverty-ridden societies. It is certainly arguable that the use of drugs such as alcohol, nicotine, or cannabis can act as at least temporary stress-relievers. There's a trade-off here that evades simplistic interpretations, and it is worth emphasising that at least 60 per cent of pregnancies that end with miscarriages or birth defects have no obvious known cause.

Focusing on these hazards may give the impression that to achieve a 'normal' birth requires a combination of good fortune and good genes, but it is precisely because of their consequences for health that these risk factors are so much better understood than those that result in that huge range of individual difference that we happily regard as representing both individuality and 'normality'. The most important aspect of all about developing systems is what may be called their built-in redundancy, meaning that there are multiple possible pathways to a successful outcome (one of the several meanings to the term 'plasticity').

There is an urgent striving towards successful living that maximises each creature's chance of surviving to make its way in the world. 'Average' or 'typical' differences quickly become submerged within the complex trajectory of individual lifelines, a process that speeds up dramatically with the drama of birth and becoming human. However, the time has not yet come to explore that question developmentally; there are some evolutionary issues to be dealt with first.

CHAPTER 4

Becoming Human

THE THREE-BILLION-YEAR SAGA OF THE EVOLUTION AND THE NINE-MONTH saga of the development of the human brain have assembled the raw materials required to begin to address the question of becoming human. Chapter 2 traced the emergence of the human brain from the origin of life itself to the appearance of *H. Sapiens*, and Chapter 3 mapped the route from the moment of conception to the birth of a baby. All species are unique, a statement that is implicit within the definition of a species; those features unique to humans include language, social existence and consciousness of self and of others. However, by contrast with other closely-related species, where there are clear continuities, these unique human facilities seem to present a sharp discontinuity with even our closest genetic and evolutionary relatives. Underlying all of them is that special attribute of humans that we call possessing a mind, of being conscious (so perhaps, *pace* George Orwell, in the case of humans some species are more unique than others!). For good reasons minds have not entered significantly into the discourse of the previous two chapters, but it is now time to begin to consider how, when and why they emerge. It is however impossible to raise such questions without also addressing the relationship between the ways in which we talk about minds and how we talk about brains.

This is not the point at which to become involved in the tortuous and voluminous philosophical debate that has surrounded what I regard as yet another spurious nineteenth-century dichotomy peculiar to Western culture, that of mind and brain. (You will surely have deduced by now that I am not very enthusiastic about dichotomies; as everyone knows, there are two types of people in the world, those who are partial to dichotomies and those who dislike them.) I would rather approach the dreaded mind theme within the evolutionary and developmental framework that the preceding chapters have established – but in doing so I will need to abandon both of the two simplest and commonest neuroscientific views about the mind/brain relationship, even though this will require me to recognise that some of my own earlier endorsements of these views have been at best over-naïve and at worst simply mistaken.[1] Neuroscientists by and large are not philosophers, but the implicit assumptions of my discipline are rather coarse: either minds are 'nothing but' the products of brains, or 'mind language' is a primitive form of what is disparagingly called folk psychology and it is time to eliminate it from scientific discourse.[2] By the time this and the following chapter reach their termination, it will I hope become clear why I now want to transcend these over-simplifications. The American poet Emily Dickinson once wrote a poem, much beloved by neuroscientists, which begins by claiming: 'The brain – is wider than the sky.' Well, I will be claiming, without resort to mysticism, and with apologies to Dickinson, that the mind is wider than the brain. Mental and conscious processes, I will insist, in the face of much contemporary philosophy, are themselves evolved and functionally-adaptive properties essential to human survival, neither descended from heaven, nor functionless add-ons, epiphenomenal consequences of having large brains, but without causal powers. My route to that conclusion will be thoroughly materialistic.

None of the essentially human attributes, of language, self- and social-consciousness that the opening paragraph of this chapter listed are present except to a rudimentary extent in other primates, nor indeed in the newborn human. Awareness, the ability to direct attention selectively to specific aspects of the environment and to be able to manipulate these aspects cognitively, is perhaps a precursor to consciousness, and awareness of this sort is surely present in many non-human species. Self-awareness may be a more complex phenomenon – one classic experiment involves painting a red spot on the forehead of a chimpanzee and then letting the animal observe itself in a mirror; chimps who have earlier seen themselves without the red spot will tend to touch and rub at it, indicating that they are capable of self-recognition. Other species do not show this capacity, but as Marc Bekoff points out,[3] not all species are so dependent on visual information and few animals have the opportunity to observe themselves – except perhaps as reflections while drinking from a pool or stream. Odour and sound are probably more relevant cues, and self-similar odour preference is shown in many species.

Thus the potential for mind must be provided by our evolutionary inheritance, its actualisation over the months and years of early childhood and the maturation of both brain and body. This is but a further illustration of the ways in which evolutionary and developmental processes and mechanisms are intimately and dynamically intertwined in the developmental system that comprises each individual's life trajectory – what I call their lifeline. The problem in discussing this dialectic is that writing – or reading – a text is inevitably a linear process, demanding an almost arbitrary separation of the two processes, and this chapter will begin where Chapter 2 left off and consider the evidence for the evolution of thoroughly modern human brains, and their relevance to human minds. However, I cannot avoid confronting the loud claims concerning the supposed evolution and genetic fixity of 'human nature' made by a newly rebranded scientific discipline once called sociobiology but now more frequently called evolutionary psychology.[4]

Hominid evolution

Leaving to one side for the moment cognitive and affective capacities, what is it that distinguishes modern humans from earlier hominids and from the great apes to whom we are so closely related genetically and evolutionarily? Modern humans differ from apes in body shape and posture, including importantly in the upright balance of the skull at the top of the spine, in bipedalism rather than knuckle-walking, in brain size relative to body weight, in possessing a longer and opposable thumb, and in our longer life span and delayed sexual maturation. How and when did these differences arise? How do they relate to the human and largely non-ape capacities for language, self-consciousness, social organisation, tool use . . . ?

Until recently the only available evidence has come from fossils, and from such human artefacts as begin to emerge over the six or so million years of evolution since the first appearance of proto-humans. As with

all matters concerning the prehistoric past, data are scarce and the interpretation contentious. Scarcely more than a couple of suitcases full of
bones comprise the evidence over which the almost legendary fights
amongst palaeoanthropologists about the dating and reconstruction of
hominid fossils have raged ever since the fossilised remains of quasi-
humans were first discovered in the Neander valley late in the nineteenth century. But this discord is as nothing compared to the mountains
of speculative claims as to the mentalities, behaviours and social organisation of our ancestors based on little more than a few shards of bones,
the fossilised remains of digested food and strangely marked stones and
skulls.

For a non-expert to step into this minefield is clearly hazardous, but
the current consensus would appear to date the divergence of pre-humans
(the earliest forms are known as hominins, the later as hominids) from
their chimpanzee cousins at some six to seven million years ago, whilst
Homo sapiens appears as distinct from other hominids around 200,000
years ago. Today, *H. sapiens* is the only surviving member of the *Homo*
genus, and it is a matter of dispute as to whether the varying other fossil
members of the genus coexisted at various times during that six-million-
year period, or whether there was only one *Homo* species at any one time,
implying some form of linear progression.[*5] Over the period, however,
there is what can be interpreted as a sequential change in the forms of
the skulls; not merely are brain sizes changing, determined from the
volume of the skull (measured as an endocast), but also the face and eye
sockets retreat under the cranium, rather than projecting forward as in
other apes. This allows for the expansion of the frontal lobes (though
the claim that humans have disproportionately expanded frontal lobes
compared with other apes, seen until recently as uncontroversial, has
been challenged).[6] The oldest skulls to be assigned *H. sapiens* status yet
discovered, in Herto, Ethiopia, have been dated at some 160,000 years
old.[7] However, to the outsider it would seem that almost every fossil
discovery implies a further early hominid species (a recent review includes
some twenty), radiating out over a period of four million years or so after
the original divergence from the ancestors of today's great apes.[8]

---

* It is too early to say how far the discovery of a miniature hominid, *H. florensiensis*, on a remote Indonesian island, which apparently existed contempor-
aneously with early humans, will affect this account.

Many of the differences in form between humans and apes emerge ontogenetically and are a reflection of the much longer developmental and maturational pathway taken by humans. We are born less mature than chimpanzees and we retain our relatively juvenile and hairless features into maturity (a phenomenon known as neoteny), leading some early evolutionists to suggest that the prime difference between us and our cousins is that we are essentially juvenile apes, a theory which still crops up from time to time.[9] However things just aren't that simple; there are differences in face size and volume, in rates of growth of different body and brain regions, and of course the skeletal changes associated with upright walking on two legs which cannot be explained by any simple theory of foetalisation.[10] The fossil record suggests that these several divergences did not appear all at once, but at different periods and in a variety of hominid species over the crucial four million years, with some longish static periods and others of relatively rapid change. For example, the enlargement in brain size in fossil hominids appears to have taken place in two distinct phases, one between 1.5 and two million years ago and one between 200,000 and 500,000 years ago, with rather little change in between. Both involve major expansions of the frontal cortex.

The more recent the hominids, the more closely they seem to have resembled modern humans in body size and shape, and indeed in brain capacity.* The earliest forms had cranial volumes of only some 320 to 380cc. The Neanderthals, who now seem to have appeared later rather than earlier than the earliest known *H. sapiens*, had larger brains – at around 1500cc they were heavier than that of many modern humans (the Herto skull volume is around 1450cc). It was perhaps this that led William Golding in his novel *The Inheritors* to ascribe much greater powers of empathy and social interaction to Neanderthals, by contrast with the more violent *sapiens* form that is supposed to have displaced them. However, it is important to remember that, for all the reasons insisted on in earlier chapters, 'the brain' is not one but a complex of organs. Thus brain size is an unreliable marker of mental and behavioural capacities, and measurements of the volume of fossil skulls[11] can tell one rather little about the internal organisation of the brain and still less about the behaviour and putative sense of agency of its owner during life.

* *H. florensiensis* may be an exception, with a rather small brain.

The genetic record

Additional, if confusing, information comes from genetics, with the oft-quoted statement that humans are genetically some 99 per cent identical to chimpanzees (the current estimate is actually 98.7 per cent). The identity is even closer amongst a numbers of genes important for brain function, although there are significant differences in the expression of these genes in the brain.[12] As no one would mistake a human for a chimpanzee phenotype, what needs to be understood, but is as yet largely shrouded in mystery, is how these minor genetic differences, presumably mainly associated with control or regulatory regions of the genome, contribute to the very different developmental cycles of two such closely related species. Much hope rests on the decoding of the chimpanzee genome – imminent as I write. However, as there is still a great deal of ignorance about the relationship of genomes to the construction of physiological form and species-typical behaviour during ontogenesis, one should not hold one's breath in the belief that revelation is soon to come.

Presumably half the present 1.1 to 1.2 per cent genetic difference between humans and chimpanzees lies in the human and half in the chimpanzee line. This, it has been calculated, will correspond to at most some 70,000 adaptively relevant amino acid substitutions in the proteins for which the DNA codes. That is really a very small number if one remembers that there are perhaps some 100,000 different proteins containing sequences amounting to around forty million amino acids present at various times and in various cells of the human body. How many of these variants are present in the brain is still unknown (although it is known that the brain expresses a greater variety of proteins than any other body organ) and still less is it clear how relevant any of them might be to any unique human brain function.

Not of course that genetic change ceased with the appearance of *H. sapiens*, despite the claims of some evolutionary psychologists (of which more below) that there has been too little time between the Pleistocene and the present day for differences to have emerged. The flow of early humans out of Africa and their steady migration across the continents has been accompanied by a variety of genetic changes,[13] often accelerated in interaction with cultural and technological developments in what has been described as gene-culture co-evolution.[14] I am squeamish about using the term evolution to refer to cultural changes because in this context it would seem to imply more than meaning simply change over time, which is the original and theory-innocent sense of the word, but rather a similarity of mechanism between cultural and biological evolution. The implication would be that cultural/social change involves a form of natural selection – an argument that has been made by a number of cultural anthropologists and sociologists.*[15] As natural selection involves differential breeding and survival as a motor of evolutionary change, it is hard to see a similar process operating in human cultures, despite the strangely popular notion of so-called 'memes' as units of culture by analogy with genes, initially proposed almost as a throwaway *bon mot* by Richard Dawkins, but now taken seriously by many who should know better.[16]

However, that cultural and technological changes are associated with genetic change is clear. To take one well-attested example, although milk is the food for all infant mammals, most mammals – and many human populations – become lactose-intolerant as they mature; the enzymes necessary to digest lactose (a major sugar component of milk) are no longer synthesised; the genes that code for them are switched off. However, with the development of agriculture and the domestication of cattle, some populations, notably those of Western Asia and Europe, developed† lactose tolerance, enabling them to digest milk and milk products later in life. A recent book by the economist Haim Ofek attributes even more significance to the development of agriculture – for with agriculture comes trading, with trading the need for symbolisation, and with symbolisation a necessary expansion of brain-processing capacity.

---

* The enthusiasm for employing out-of-place Darwinian metaphors is one of the less attractive features of much current writing in the area – as when in his book *Darwin's Dangerous Idea*, philosopher Daniel Dennett refers to natural selection as 'a universal acid', a mechanism for change that runs from subatomic physics to music and the arts.

† Using the word 'developed' here masks possible mechanisms. One explanation, the one most acceptable to strict neo-Darwinians, would be that some individuals in the generally lactose-intolerant population acquired a mutation in which the lactose synthesis was no longer switched off during development, and that this conveyed a selective advantage, by enabling them as adults to digest food that the less fortunate members of the population could not. There are other ways of explaining what otherwise appears as a 'directed' form of evolutionary change, but to explore them here would take me too far out of the mainstream of my argument.

For Ofek, the first symbol was money and its invention was a driving force in human evolution.[17]

Evolving human minds

The crucial questions that evolutionary prehistory, palaeoanthropology and archaeology are called upon to answer are at what stage in the transition from hominins to hominids to humans do species-specific human capacities emerge? Should we attribute human-ness to all of these? How far do the manufacture and use of tools or fire or the care with which the dead are buried imply the existence of language? And from the perspective of this book, what do brains have to do with any of these? It seems pretty obvious that all of these capacities interact; it is difficult to imagine culture, human social organisation or the spread of new technologies without language, or the possession of language without a concept of self, or what some theorists have referred to as a 'theory of mind'[18] – and the argument about which came first, which has vexed some theorists, is pretty sterile. That our human brains and minds are products of evolution is unequivocal, but the processes whereby modern minds have evolved, and the constraints that these evolutionary processes may have imposed are both matters of intense and passionate debate.

In Chapter 2 I pointed out that in studying the evolutionary record, it often became necessary to treat contemporary species as if they were representatives of past, now extinct species, even though all modern forms have evolved *pari passu* with humans. This inference of the past from the present enabled me to speak of present-day fish or amphibian or mammalian brains as indicators of the path of evolution of human brains – a seductive but problematic approach. The same is true in attempting to understand human mental origins. The comparative method is difficult, as there is only one living species of *Homo* – the *sapiens* form. We can, as I did above and will return to below, compare ourselves with *Pan troglodytes*, the chimpanzee. And we can look for differences and similarities amongst living human populations, the stock-in-trade of European and US anthropologists over the last couple of centuries. The problem of course is that all contemporary human societies, whether modern European, Trobriand Islanders or First Americans, are the products of 200,000-odd years of genetic and cultural change, of adaptation to particular local climatic and geographical circumstance.

Over those two centuries of Euro-American research, the history of the sciences attempting to obtain a purchase on interpreting human nature has oscillated between two poles. On the one hand there have been those arguing that the differences between human populations (so-called 'races') are so great as to imply significant average genetic differences between them, accounting for alleged differences in 'intelligence' and other cultural and behavioural features. The history of this form of scientific racism has been told many times[19], and if we set to one side the neo-Nazi political parties endeavouring to draw sustenance from it, and those few academics who seek notoriety by pandering to them, it may safely be discounted. Modern genetic evidence demonstrates that whilst there are average genetic differences between human populations, these do not map on to socially ascribed 'racial' divisions. Thus no racial distinction is drawn between the populations of north and south Wales, yet there are small differences in average genetic profile between the two populations. And conversely, genetically Polish Jews resemble their Catholic neighbours more closely than they do Jews from, for instance, Morocco. In fact the overwhelming proportion of the genetic difference between individuals lies within rather than between so-called races – leading most modern population biologists to discard as unhelpful the term 'race' in the human context.* The more useful, if ponderous, term is biogeographical diversity.

The alternative to attempting to root human *differences* in biology is to seek for alleged human *universals*. If there are such universals, which distinguish members of the species *Homo sapiens* from *Pan troglodytes*, then, it is argued, they must be the products of evolution, either directly adaptive to the human way of life, or an epiphenomenal consequence of such adaptation. The identification of such universals has become

---

* Of course this is not to discount the still huge social, political and ideological power of the concept of race. Racism, based on differences or alleged differences in skin colour, ethnic origin, language or culture is still a poisonous source of hate in the world. And some of these differences are defined by biology – for example, skin colour identifies a 'black' person as different in a society composed primarily of an indigenous 'white' population. Furthermore, as Nancy Stepan has pointed out (*The Idea of Race in Science*, Macmillan, 1982), racism is a 'scavenger ideology' – it will pick up and employ out of context and in its favour any current scientific hypothesis.

the province of a relatively newly emerged science, that of evolutionary psychology. The declared aim of evolutionary psychology is to provide explanations for the patterns of human activity and the forms of organisation of human society which take into account the fact that humans are evolved animals. To state the obvious, the facts that we humans are bipedal, live for some sixty-plus years, that we have offspring that are born relatively helpless and need long periods of caregiving before they are able to survive autonomously, that we have limited sensory, visual and auditory ranges and communicate by language, have shaped and constrained the way we think (our 'psychology' therefore) and the forms of society we create. An integrated biosocial perspective is thus an essential framework within which to attempt any understanding of 'human nature'. So far, so uncontroversial. The problem, about which I have written extensively elsewhere,[20] is that, like their predecessor sociobiologists, a group of highly articulate over-zealous theorists* have hijacked the term evolutionary psychology and employed it to offer yet another reductionist account in which presumed genetic and evolutionary explanations imperialise and attempt to replace all others.

Their central claim is that 'becoming human' occurred at some point in an assumed 'environment of evolutionary adaptation' or EEA, some time in the Pleistocene, anywhere from 600,000 to 100,000 years ago, but most likely around the time assigned to the Herto skull, in which crucial universal aspects of human nature adapted to social living became genetically 'fixed', so that even where these forms of behaviour are currently maladaptive, there has not been sufficient evolutionary time subsequently to modify them to adapt to the conditions of modern life. Just what were the evolutionary/environmental pressures that led to the emergence of human minds and consciousness are a matter of speculation – one theory has it that this period, beginning about 250,000 years ago, showed substantial climatic variations in Africa, with rapid transitions between hot and humid and cold and dry, which posed huge survival problems, forcing the selection of the mental skills capable of withstanding them – adapt or go extinct.

---

* Prominent members of this group, largely north American, include social scientists Leda Cosmides and John Tooby, psychologists Martin Daly and Margo Wilson, cognitive psychologist Steven Pinker and animal behaviourists Randy Thornhill and Craig Palmer. Surrounding them are a number of over-impressed philosophers and UK-based acolytes.

Now although the Pleistocene may well have been a crucial period for the emergence of specifically human attributes, the assumption of 'fixity' misspeaks and impoverishes modern biology's understanding of living systems, in three key areas: the processes of evolution, of development and of neural function. Underlying all of these are two major conceptual errors: the misunderstanding of the relationship between enabling and causal mechanisms, and the attempt to privilege distal over proximal causes. It is on these shaky foundations that prescriptions for how humans do and must behave (and often too for the social policies that flow from this) are based.

Evolutionary psychologists go to some lengths to insist that, unlike exponents of earlier versions of social Darwinism, they are not genetic determinists, or as they sometimes put it, nativists. Rather, they argue, as indeed would most modern biological and social scientists (with the possible exception of psychometricians and behaviour geneticists) that the nature/nurture dichotomy is a fallacious one. Indeed, they are often at pains to distinguish themselves from behaviour geneticists, and there is some hostility between the two.[21] Instead they seek to account for what they believe to be universals in terms of a version of Darwinian theory which gives primacy to genic explanations. For evolutionary psychology, minds are thus merely surrogate mechanisms by which the naked replicators whose emergence I described in Chapter 2 enhance their fitness. Brains and minds have evolved for a single purpose, sex, as the neuroscientist Michael Gazzaniga rather bluntly puts it.[22] And yet in practice evolutionary psychology theorists, who as Hilary Rose points out,[23] are not themselves neuroscientists or even, by and large, biologists, show as great a disdain for relating their theoretical constructs to real brains as did the now discredited behaviourist psychologists they so despise.

Particularly in its attempt to account for such seemingly non-Darwinian human characteristics as self-sacrifice and altruism – actions apparently not in 'the interests' of our genes – evolutionary psychology draws heavily on the theory of 'kin selection' based on 'inclusive fitness', ideas developed in the 1960s by William Hamilton as logical development of Darwinian natural selection. Fitness is a measure of the extent to which particular forms or combinations of genes are passed from one generation to the next. Put crudely, the fitter you are, according to this very specific use of the term, the more the genes you possess will crop up in the next generation (it is thus a tautological definition, as pointed out many years ago by Stephen Jay Gould). Inclusive fitness broadens the concept by recognising that as you share genes in common with your relatives, you can assist their transmission to the next generation by increasing the reproductive success of your relatives. In developing these ideas, Hamilton in essence formalised an often recounted pub comment by the geneticist JBS Haldane in the 1950s, that he would be prepared to sacrifice his life for two brothers or eight cousins. Because he shared genes in common with his kin, the proportion varying with the closeness of the kinship, then 'his' genes would serve their interest in replication not only by ensuring that he himself had offspring (which Haldane himself did not), but by aiding in the replication of the identical copies of themselves present in his relatives.[24]

This is kin selection, and became the core theory of EO Wilson's 'new synthesis' of sociobiology, which by the 1990s had mutated into evolutionary psychology. Within this theoretical framework, what Steven Pinker calls the 'architecture' of our minds, and our forms of social organisation, must be seen as adaptations shaped by natural selection to achieve the goals of the optimal replication of the genes of individual humans and their genetic close relatives, that is, an increase in inclusive fitness as predicted by kin selection. Thus apparently disinterested human behaviour is 'in reality' a way of enhancing genetic success, either directly by increasing our social prestige and hence reproductive attractiveness to potential sexual partners, or indirectly by improving the life chances of those to whom we are genetically related. In the cases in which it is hard to argue that altruistic behaviour can be reduced to such genetic pay-offs, an alternative mechanism, called 'reciprocal altruism' ('you scratch my back and I'll scratch yours') has been advanced by Robert Trivers.[25]

It is here that, despite their disavowals, the nativism of the evolutionary psychology enthusiasts becomes clear. When the anthropologist Laura Betzig writes[26] that everything from pregnancy complications to the attraction of money – to say nothing of female coyness, male aggression, and, in Randy Thornhill and Craig Palmer's hands, rape[27] – can be explained by invoking Darwinian mechanisms, the conclusion is inescapable: genetic mechanisms, selected by evolutionary adaptation and fixed in the EEA, underlie these human problems and proclivities. More sophisticated evolutionary psychologists finesse the argument by maintaining that they are not really saying that there are specific actual genes – lengths of DNA – for being attracted sexually by symmetrical bodies, or disliking spinach whilst a child but liking it as an adult, to take but two of the explanatory claims of evolutionary psychology. Instead there are mechanisms, ultimately encoded in the genes, which ensure that we *are*, on average, so attracted, albeit the way in which the genes ensure this is mediated by more proximal mechanisms – by, for instance, creating modular minds whose architecture predisposes this type of behaviour. The spread of such theoretical pseudogenes, and their putative effect on inclusive fitness can then be modelled satisfactorily *as if* they existed, without the need to touch empirical biological ground at all.

Evolutionary time

A further characteristic feature of the evolutionary psychology argument is to point to the relatively short period, in geological and evolutionary terms, over which *Homo sapiens* – and in particular modern society – has appeared. Forms of behaviour or social organisation which evolved adaptively over many generations in human hunter-gatherer society may or may not be adaptive in modern industrial society, but have, it is claimed, become to a degree fixed by humanity's evolutionary experience in the hypothetical EEA. Hence they are now relatively unmodifiable, even if dysfunctional. There are two troubles with such claims. The first is that the descriptions that evolutionary psychology offers of what human hunter-gatherer societies were like read little better than Just-So accounts, rather like those museum – and cartoon – montages of hunter-dad bringing home the meat whilst gatherer-mum tends the fireplace and kids, so neatly decoded by Donna Haraway in *Primate Visions*.[28] There is a circularity about reading this version of the present into the past – what Hilary Rose and I have elsewhere called Flintstone psychology – and then claiming that this imagined past explains the present.

However, the more fundamental point is the assertion by evolutionary psychologists that the timescale of human history has been too short for evolutionary selection pressures to have produced significant change. The problem with this is that we know very little about just how fast such change can occur. Allowing fifteen to twenty years as a generation time, there have been as many as 11,000 generations between the Herto fossils and the present. Whilst it is possible to calculate mutation rates and hence potential rates of genetic change, such rates do not 'read off' simply into rates of phenotypic change. As Stephen Jay Gould and Niles Eldredge have pointed out in developing their theory of punctuated equilibrium, the fossil record shows periods of many millions of years of apparent phenotypic stasis, punctuated by relatively brief periods of rapid change.[29] This is simply because the many levels of mediation between genes and phenotype mean that genetic change can accumulate slowly until at a critical point it becomes canalised into rapid and substantial phenotypic change.

A 'Darwin' is the term used to provide a measure of the rate of evolutionary change. It is based on how the average proportional size of any feature alters over the years and is defined as 1 unit per million years. Laboratory and field experiments in species varying from fruit-flies to guppies give rates of change of up to 50,000 Darwins. Steve Jones describes how English sparrows transported to the south of the US have lengthened their legs at a rate of around 100,000 Darwins, or 5 per cent a century.[30] So we really have no idea whether the 11,000 or so generations between Herto and modern humans is 'time enough' for substantial evolutionary change. We don't even know what 'substantial' might mean in this context; is the evolution of adult lactose tolerance substantial? It has certainly made possible – and been made possible by – very different styles of living from those of earlier hunter-gatherer societies, recognised even in such biblical accounts as that of Cain and Abel. However, granted the very rapid changes in human environment, social organisation, technology and mode of production that have clearly occurred over that period, one must assume significant selection pressures operating. It would be interesting in this context to calculate the spread of myopia – which is at least in part heritable, and must in human past have been selected against – once the technological and social developments occurred which have made the almost universal provision of vision-correcting glasses available in industrial societies. What is clear, however, is that the automatic assumption that the Pleistocene was an EEA in which fundamental human traits were fixed in our hominid ancestors, and that there has not been time since to alter them, does not bear serious inspection.

'Architectural' minds

Unlike earlier generations of genetic determinists, evolutionary psychologists argue that these proximal processes are not so much the direct product of gene action, but of the evolutionary sculpting of the human mind. The argument, drawing heavily on the jargon and conceptual framework of artificial intelligence, goes as follows: the mind is a cognitive machine, an information-processing device instantiated in the brain. But it is not a general-purpose computer; rather it is composed of a number of specific modules (for instance, a speech module, a number-sense module,[31] a face-recognition module, a cheat-detector module, and so forth). The case for the modularity of mind was first advanced in 1983 by the psychologist Jerry Fodor in an influential book[32] whose claims will continue to occupy me in later chapters. Steven Pinker's claim, however, goes further: the modules, he argues, have evolved quasi-independently during the evolution of early humanity, and have persisted unmodified throughout historical time, underlying the proximal mechanisms that traditional psychology describes in terms of motivation, drive, attention and so forth. The archaeologist Steven Mithen, by contrast, makes a precisely converse argument, claiming that whereas in non-human species the modules were and are indeed distinct, the characteristic of human mentality is the ability to integrate, so that rather than functioning in modular manner based on the presence of a toolkit of quasi-autonomous specialised functional units, like a Swiss army knife (the favoured analogy of Tooby and Cosmides), the human brain is like a general-purpose computer.[33]

In any event, whether such modules, if they exist, are more than theoretical entities is unclear, at least to most neuroscientists. Indeed Pinker goes to some lengths to make it clear that the 'mental modules' he invents do not, or at least do not necessarily, map on to specific brain structures.* But, as will become clear in the argument from

---

* The insistence of evolutionary psychology theorists on modularity puts a particular strain on their otherwise heaven-made alliance with behaviour geneticists. For instance IQ theorists, such as the psychometrician Robert Plomin, are committed to the view that intelligence, far from being modular, can be reduced to a single underlying factor, g, or 'crystallised intelligence' (Plomin, R, Owen, MJ and McGuffin, P, 'The genetic basis of complex human behaviors', *Science*, 264, 1994, 1733–7). A similar position was emphatically taken by Herrnstein and

development that follows in the next chapter, even if mental modules do exist, they can as well be acquired as innate. Fodor himself has taken some pains to dissociate himself from his seeming followers, in a highly critical review of Pinker, culminating in a book with the intriguing title *The Mind Doesn't Work That Way.*[34] Mental modules there may be, this revisionist text argues, but only for lower-level activities, not for those involving complex cognitive and affective engagement.

Modules or no, it is not adequate to reduce the mind/brain to nothing more than a cognitive, 'architectural' information-processing machine. As I continue to insist throughout this book, brains/minds do not just deal with information. They are concerned with living meaning.[35] In *How the Mind Works* Pinker offers the example of a footprint as conveying information. My response is to think of Robinson Crusoe on his island, finding a footprint in the sand. First he has to interpret the mark in the sand as that of a foot, and recognise that it is not his own. But what does it mean to him? Pleasure at the prospect of, at last, another human being to talk and interact with? Fear that this human may be dangerous? Memories of the social life of which he has been deprived for many years? There is a turmoil of thoughts and emotions within which the visual information conveyed by the footprint is embedded. The key here is emotion, for the key feature which distinguishes brains/minds from computers is their/our capacity to experience emotion and express feelings. Indeed, emotion is primary, even if by that term we mean something less crudely reductionist than Gazzaniga's three-letter simplification. Perhaps this is why Darwin devoted an entire book to emotion rather than to cognition.

Emotions are evolved properties, and several neuroscientists have devoted considerable attention to the mechanisms and survival advantages of emotion.[36] Antonio Damasio, for example, distinguishes between emotions that are physiological phenomena possessed to a degree by all living organisms, as I mentioned in Chapter 2, and feelings, the mental states that are associated with emotions, that are unique to humans. So it is, therefore, all the more surprising to find this conspicuous gap in the concerns of evolutionary psychologists – but perhaps this is because

---

Murray (Herrnstein, RJ and Murray, C, *The Bell Curve*, Simon and Schuster, New York, 1996) who argue that whatever intelligence is, it cannot be dissociated into modules!

not even they can speak of a 'module' for feeling, or even emotion. Rather, affect and cognition are inextricably engaged in all brain and mind processes, creating meaning out of information – just one more reason why brains aren't computers. What is particularly egregious in this context is the phrase, repeated frequently by Leda Cosmides, John Tooby and their followers, 'the architecture of the mind'. Architecture, which implies static structure, built to blueprints and thereafter stable, could not be a more inappropriate way to view the fluid dynamic processes whereby our minds/brains develop and create order out of the blooming buzzing confusion of the world which confronts us moment by moment.

The problem is indicated even more sharply by the reprinting of a series of classical anthropological and sociobiological papers in the collection entitled *Human Nature*[37]. The editor Laura Betzig's view is that these show the way that Darwinian insights transform our understanding of social organisation. The papers were largely published in the 1970s and 1980s, and for their republication in 1997 each author was asked to reflect in retrospect on their findings. What is interesting is that when the anthropologists go back to their field subjects, the subjects report rapid changes in their styles of living. Kipsigis women no longer prefer wealthy men (Borgerhoff, Mulder), the Yanonomo are no longer as violent as in the past (Chagnon), wealth no longer predicts the number of children reared (Gaulin and Boster) and so forth. Each of these societies has undergone rapid economic, technological and social change in the last decade. What has happened to the evolutionary psychology predictions? Why have these assumed human universals suddenly failed to operate? Has there been a sudden increase in mutation rates? Have the peoples they had studied taken Richard Dawkins to heart and decided to rebel against the tyranny of their selfish replicators?

Enabling versus causing; proximal versus distal explanations

Once the inflated claims of evolutionary psychology have been discarded, what is left of human universals and their emergence during evolution? It is important to distinguish between evolutionary processes that *enable* modern human thought and action, and those that apparently *determine* it, as implied by much evolutionary speculation. To take another example from the literature of evolutionary psychology, Pinker, in *How the Mind Works*, claims that (with the engaging exception of what he describes as 'great art') humans show a universal propensity to prefer pictures containing green landscapes and water, so-called Bayswater Road art.[38] He speculates that this preference may have arisen during human evolution in the EEA of the African savannah. The grander such assertions, the flimsier and more anecdotal becomes the evidence on which they are based. Has Pinker ever seen savannah, one wonders? Is this so-called universal preference shared by Inuits, Bedouins, Amazonian tribespeople? Or is it, like so much research in the psychological literature, based on the samples most readily available to US academics – their own undergraduate students? It is hard not to be reminded of the argument once made by an ophthalmologist that El Greco's characteristic elongated figures were the result of his astigmatism.

The point is that there are much simpler proximal explanations for such preferences should they occur – that in Western urban societies, as Simon Schama points out, 'the countryside' and 'the wilderness' have become associated with particular Arcadian and mythic qualities of escape from life's more pressing problems.[39] Such proximal mechanisms, which relate to human development, history and culture, are much more evidence-based as determining levels of causation, should these be required, than evolutionary speculations. It is surely an essential feature of effective science and of useful explanation to find an appropriate – determining – level for the phenomenon one wishes to discuss. As an example, consider the flurry of attempts, particularly in the US, to 'explain' violent crime by seeking abnormal genes or disordered biochemistry, rather than observing the very different rates of homicide by firearms between, say, the US and Europe, or even within the US over time, and relating these to the number and availability of handguns.[40] Despite the implicit and sometimes even explicit claim that if we can find an evolutionary explanation for a phenomenon, this will help us fix it,[41] it seems highly doubtful that evolutionary psychology or behaviour genetics will ever contribute anything useful to either art appreciation or crime prevention. The enthusiasm in such cases for proposing biological causal mechanisms owes more to the fervour of the ultra-Darwinians to achieve what EO Wilson calls consilience[42] than to serious scholarship.

What a more modest understanding of evolutionary processes can lead us to recognise is that the evolutionary path that leads to humans has produced organisms with highly plastic, adaptable, conscious brains/minds and ways of living. By virtue of our large brains, and presumably the properties of mind and consciousness that these brains permit, humans have created societies, invented technologies and cultures, and in doing so have changed themselves, their states of consciousness and, in effect, their genes. We are the inheritors of not merely the genes, but also the cultures and technologies of our forebears. We have been profoundly shaped by them and, by being so shaped, can in turn help shape our own futures and those of our offspring.

The evolution of language

Even without the inflated claims of evolutionary psychologists, there are some hard questions to be asked about the emergence of human species-specific attributes and behaviours. It is difficult to imagine effective social organisation or the spread of new technologies without language, or the possession of language without a concept of self, and at least a rudimentary theory of mind. Are minds possible without language, or language without minds? It would seem that neither can do without the other any more than proteins can do without nucleic acids and vice versa. The debate about which came first may seem rather sterile, but as minds are notoriously hard to define and language relatively easy there is a sense in focusing on the latter.

Interspecific communication has of course a long evolutionary history, whether by chemical (pheromonal) or visual signals. Even in asocial species sexual reproduction means that males and females must find one another and signal willingness or otherwise to mate. Auditory signalling must have arisen with the emergence of land animals. The cacophony of croaking frogs and the morning chorus of birdsong around my rural retreat in April and May are obvious examples. With the appearance of social species communication becomes more sophisticated, enabling the presence of prey or predator to be signalled. Amongst the most elaborated are the warning calls of chimpanzees, who can signal not merely the presence of predators but whether they are aerial (eagle) or terrestrial (snake).[43]

In a stimulating move into the biological terrain, the linguist Noam Chomsky, writing with the ethologist Marc Hauser and Tecumseh Fitch, has pointed to a peculiar difference between the universality of the genetic mechanism based on DNA and the species-uniqueness of all such communication systems, which are seemingly meaningful only to conspecifics.[44] Furthermore, non-human communication systems, unlike human speech, are closed; as Hauser and his colleagues put it, they lack the rich expressive and open-ended power of human language. Even the gestural communications of chimpanzees and gorillas are essentially signals, not symbols. Communicative signals may be systematic and coded to support a variety of different instructions or intentions, but they are used strictly as indices, whereas symbols demand intersubjectivity. Symbols are conventional, resting on a shared understanding that any particular symbol is a token representing a particular feature of a particular referential class. Unlike a signal, which can be regarded as an instruction to behave in a particular way, a symbol guides not behaviour but understanding.[45] Furthermore, signals are not recursive – that is, they cannot be used to relate to anything but the immediate context and situation; they cannot reflect or elaborate upon it. Human language capacity is by contrast essentially infinite – there is no obvious limit to the types of meaningful sentences that can be constructed, and this distinguishes it from all known non-human forms of communication.

So when do these abilities arise in human evolution? Hominid practices including the making and use of tools such as flint hammers and axes, the use of fire, and ritual burial of the dead, predate the appearance of *H. sapiens*. Does this imply that other hominids too could speak, and utilised language? If so, or indeed if not so, how and when did language emerge? To be able to communicate in speech requires two distinct capacities on the part of the speaker. One is essentially mechanical: to have the relevant muscular and neural control over vocalisation, the larynx, mouth and tongue need to be able to articulate a variety of distinctive sounds, a skill which human children need to develop as they learn to speak. Vocalisation is dependent on midbrain structures, especially a midbrain region close to the ventricles, rich in hormone receptors and in communication with amygdala and hippocampus, a region known as the periaqueductal grey. Terrence Deacon has argued that the move from generalised vocalisation to the ability to speak is related to an upwards shift in hominid evolution during which this region comes increasingly under cortical control.[46] Granted that this has already occurred in chimpanzees, and that their control over tongue and facial movements and their neural representation is pretty similar to those in humans, and they have adequate auditory perceptive skills, why can't – or won't – chimpanzees speak – or even imitate human speech as can parrots? The primatologist Sue Savage-Rumbaugh argues that this is a mere mechanical problem: because of the vertical position of the human head above the spinal column, our vocal tract curves downwards at a 90-degree angle at the point where the oral and pharyngeal cavities meet, whereas the ape vocal tract slopes gently downwards. This means that in humans the larynx and tongue are lower and the nasal cavity can be closed off, enabling the possibility of lower pitched vowel-like sounds.[47]

The second requirement for linguistic communication is cognitive: the ability to use vocalisations to symbolise objects, processes, attributes; that is, to use speech to communicate in language. This is turn requires that the receiver of the communication can distinguish the vocalisations, and shares a common understanding of their referents in the external world, which minimally depends on the relevant auditory perceptual apparatus making it possible to hear and distinguish sounds. But it is also necessary to possess the power to compute and categorise these sounds, an ability that must depend on brain structures distinct from those involved in the mechanics of sound production. Modern human brains are asymmetric; the left hemisphere regions known for their discoverers as Broca's area and Wernicke's area are noticeably larger than the corresponding right hemisphere regions. These regions are of vital importance in the generation of coherent speech. The French neurologist Paul Broca was the first modern researcher unequivocally to allocate a function to a brain region based on a lesion in that region in a patient he studied who had lost the power of speech and whose vocabulary was essentially limited to the sound 'tan'. After the patient's death the autopsy showed a left hemisphere lesion, in a brain region not far above the left ear in life, which has ever since been known as Broca's area. More posteriorly lies Wernicke's area, while between them is the primary auditory cortex. However, this asymmetry is apparent not merely in the hominid endocasts, but also in chimpanzees, which suggests that the brain regions required for speech were already present prior to its emergence.

Nor is a genetic approach more helpful. For instance, there was a great fanfare a few years back when a so-called 'language gene' (coding for a protein called FOXP2, which acts to regulate the expression of other genes) had been discovered.[48] Members of a human family carrying a mutation in this gene suffer from a severe language and speech disorder, and the press hailed the discovery as indicating that a genetic basis for human linguistic capacity had been discovered, although this interpretation was soon recognised as much oversimplified.[49] But it also turns out that FOXP2, differing only in a couple of amino acids in its protein sequence, is also present in gorillas and chimpanzees and, with a few more variations, in mice. Whether these minor variations are enough to contribute to the profound difference in linguistic capacity between mice, chimps and humans remains to be resolved, but it does not seem *a priori* likely. It must be assumed that many of the genetic changes relevant to constructing a chimpanzee rather than a human phenotype are likely to lie in complex regulatory mechanisms rather than in the structural proteins which comprise the bulk of both chimpanzee and human brains and bodies. And once again the apparently genetic question becomes instead a developmental one.

If, despite possessing the relevant genes and brain structures, chimpanzees can't speak, it certainly isn't for want of humans attempting to teach them. For several decades now researchers have been heroically taking infant chimpanzees into their homes and lives and attempting to teach them, not speech, but the logical manipulation of symbols, cards or computer graphics, associated with specific nouns and verbs – banana, water, want, fetch, tickle and so forth. The most devoted perhaps have been Sue Savage-Rumbaugh and Duane Rumbaugh, whose most successful protégé has been a pygmy chimp (a bonobo) called Kanzi.[50] After years of training and hundreds of thousands of trials Kanzi can go further than merely manipulating symbols using a keyboard; he can apparently also respond to normal spoken English by, for example, correctly interpreting requests such as 'Put the soap on the apple.' Kanzi was raised by a foster-mother who was the initial subject for the training experiment, but he quickly surpassed her in 'linguistic' ability, perhaps, as Deacon suggests, because he was immature at the time that his mother was being taught, and implicitly acquired the skills she was supposed to be explicitly learning. But if developing language is dependent on some critical* period of maturation of the brain then why don't infant chimpanzees in the wild develop language?

Despite Kanzi's skills, his vocabulary and use of it are strictly limited, and of course he cannot use it to communicate with other chimpanzees, only with his human tutors. Savage-Rumbaugh insists that it is simply his lack of ability to vocalise; that he understands and responds appropriately to her own spoken speech, and will initiate actions and conversations by grimacing, pointing, and using his lexigram: boards full of manipulable symbols that chimpanzee speech-trainers have developed. Kanzi thus has in this sense a theory of mind. He can recognise himself in a mirror, as can other chimpanzees, and he can distinguish by his behaviour the intentions of other chimpanzees (even rhesus monkeys show some capacities in this direction[51]). Furthermore he is capable of seemingly co-operative or even altruistic behaviour towards his fellows. Kanzi's capacities, as expressed in interaction with Savage-Rumbaugh and her devoted collaborators, imply an evolutionary preparedness, or, in Gould's terms, the structures and properties available for an exaptation. Speech thus only awaits the human context, and the fortuitous alterations in the position of the human larynx, itself a consequence of upright bipedal locomotion rather than knuckle-walking as employed by our fellow apes. And why, when humans 'came down from the trees' in that ancient savannah, did our ancestors turn from knuckle-walking to full bipedalism? So that they could see above the height of the grass, some have argued. Or so as to diminish the area of body exposed to full sun, argue others. If this is so, human language may well be an exaptation, though one that has profoundly shaped the course of humanity's subsequent evolution.

Hauser and his colleagues argue the need to distinguish between linguistic ability in the broad sense, including all the biological apparatus required to perform it, and in the narrow sense, corresponding to

---

* Biologists concerned with development and behaviour sometimes speak of critical periods as if there are very tight time windows during which particular structures or behaviours are formed. However, especially in the context of behavioural competences, it might be more appropriate to refer to sensitive rather than critical periods, to emphasise, as Pat Bateson does, the degrees of flexibility that are available.

what they call an internal abstract computational system that generates internal representations (mental language) and maps them on to sensory-motor outputs, although they offer no thoughts as to which specific brain regions might be involved in these processes. Does Kanzi have such an abstract computational system embedded in his bonobo brain, as perhaps demonstrated by his capacity to use symbols to signal needs? If so, then language evolved slowly and incrementally through hominid evolution. If not – and they think not – then it arrived as one giant step for hominidkind relatively recently.

If evolutionary comparisons cannot answer the question of the origins of human linguistic capabilities, perhaps developmental studies can? I'll come back to the empirical evidence in the next chapter, against the background of the revolution in linguistic theory associated with the name of Noam Chomsky from the 1950s. He insisted that, far from being acquired developmentally, language depends on an innate 'grammar module' in the mind which ensures that all human languages are constructed upon a universal grammar, and that this represents a qualitative distinction between humans and all other life forms, a unique ability.[52]

Chomsky appears at that period to have been indifferent to the biological mechanisms involved in this seemingly innate structure; it was left to Pinker, his one-time pupil, to seemingly disassociate himself from his earlier mentor by arguing for what he called 'the language instinct', still innate but now embedded in the brain, as an evolved biological phenomenon.[53] In the computer-speak that Pinker favours, humans possess a 'language acquisition device' or LAD. However, it would appear that, as is the case with other such proposed modules, if it exists, it has no particular instantiation within brain structures or biochemistry, although Broca's and Wernicke's areas are essential components of any language acquisition system. As I pointed out above, Pinker's 'modules' are mental – or perhaps merely metaphorical. Chimpanzees presumably do not possess such a LAD, or they would acquire language in the wild – but if Kanzi can acquire some form of language if he is taught at an early enough stage in development and only lacks the mechanical structures required to express it, then the LAD cannot be an innate module, and linguistic capacity must be acquired during some critical developmental/maturational period.

The disputes over a saltationist versus gradualist view of the evolution of human language and of its innate versus developmental acquisition have been intense. Although the two issues would seem in principle separable, in practice the rival theories are linked – if, like Chomsky and Pinker, you are an innatist you are likely also to be a saltationist, and reciprocally, gradualism and developmentalism go together. My own take on the debate must therefore wait until the next chapter, in which I turn from an evolutionary to the developmental perspective. For now I want to remain with the former. The most systematic anti-saltationist/innatist arguments have come from Terrence Deacon, on whose work I drew earlier, and from psychologist Merlin Donald in his book *A Mind so Rare*.[54] For them, as indeed for me, language is inseparable from culture, and culture implies a recognition that there are individuals in the world other than oneself who resemble oneself in having a mind, intentionality, agency. If we did not attribute minds similar to our own to our fellow humans, there would be little point in endeavouring to communicate with them, symbolically or linguistically. As Deacon points out, language involves symbolic manipulation, whether the symbols are sounds, artefacts or computer generated as they are for Kanzi. Symbolic representation appears early in human culture. It is a crucial accompaniment to social group living, tool use, and the division of labour that larger social group sizes permit (for instance, between hunting and gathering, care-giving and exploring new living sites). Witness the cave paintings, the earliest of which, found in France and Spain, have been dated back as far as 50,000 years ago. Indeed some argue an even earlier date for red ochre engraved with geometrical designs found in Blombos cave in South Africa and dated at 77,000 years ago. Even prior to that, the Herto skulls from Ethiopia, referred to earlier, show signs of ritual cuttings that suggest that even 160,000 years ago, our ancestors recognised the common humanity of their dead. Nor does the ancient art of Africa or Australia, or the rich cave art of France and Spain, with its abstract decorative hand patterns, stick-people and hunting parties, its astonishingly realistic deer and bison, resemble in any way Pinker's Bayswater Road fantasies.

Humans, like our ape cousins, are social animals, and social living requires communication. Chimp communication is limited by their physiology and anatomy; conversely human speech is made possible by ours. Thus the descent of the larynx enables rich auditory communication; the possibility itself makes feasible more complex forms of group organisation and life styles, and this in turn encourages the expansion of language. The expansion of language and symbolic communication itself represents a selection pressure on those brain areas (Broca's and Wernicke's regions, the frontal lobes) that are necessary to articulate and interpret it, and the bootstrapping process associated with the great expansion of hominid brains between 500,000 and 200,000 years ago results. Language and symbolic representation develop *pari passu* with culture and social organisation, as the most efficient way of ensuring understanding between members of the same family or group. Thus brain, language, and social group living co-evolve. There is no LAD, no Chomskian pre-ordained universal grammar; rather, language evolves to fit comfortably with pre-existing brain mechanisms, and the brain evolves to accommodate these new linguistic potentials. And, as Mithen argues, even if non-human brains and minds were more strictly modular, the very appearance of language demodularises mind.

What must be common ground is that, however they evolved, language and symbolic representation and manipulation became central features of human social organisation, and in doing so provided the motor for all subsequent cultural, technical and genetic change. Once established, 200,000 or so years ago, there was no turning back for *H. sapiens*. The paths that have led to our many contemporary cultures, however much they may have diverged since or are now being reunified in a new globalisation, began here, in humanity's long march out of Africa.

CHAPTER 5

Becoming a Person

Babies and persons

IS A NEWBORN BABY ALREADY HUMAN? YES, BUT NOT QUITE. IN IMPORTANT ways she is not yet a person with independent agency, but a pre-human in the process of becoming a person. Like many other infant mammals, the infant human is born only half-hatched, altricial. Several years of post-natal development – more than any other species – are needed before a person, let alone a mature person, begins to emerge, and these years demand the engagement of caregivers. We accord babies human rights, though not yet human duties. Newborns cannot be held responsible for their actions; we do not regard them as emerging from the womb with a fully-fledged theory of mind. The age at which such responsibilities are assumed varies from culture to culture, time to time and context to context; poor agricultural economies may send their children out to tend flocks by the time they are six, and Victorian England was putting youngsters not much older up chimneys. Puberty is in many cultures taken as a symbolic transition point between child and adult, yet the notorious child soldiers composing the raggle-taggle armies fighting for control in many parts of Africa are often younger, and Britain still permits boy soldiers to join the army before they are legally entitled to vote.

Enabled by and enabling these differences in social practice, the acquisition of consciousness, language, culture, all occur within the first few post-natal years of the ontogeny of brain and body. It is over this extended period that human brains go on developing and maturing; nerves are becoming myelinated, thus speeding communication, more neurons are being born in key regions of the cortex, and new synaptic connections are being made. The autopoietic processes by which a newborn achieves personhood, exploiting the resources of its genes and environment, and above all the social interaction with caregivers, are central to our understanding of what it is to be human.

On instinct

For 'normal' babies brought up in a 'normal' range of environments, physical and mental development proceeds along a seemingly seamless trajectory, in a well-charted sequence in which smiling, walking, speech, self-toileting and so on appear over the first two years of life. It is easy to see this as the unrolling of some genetically pre-ordained pattern, and to speak of behaviour as being 'innate' or 'instinctive'. Yet such easy descriptors are misleading. In its origin, the word 'innate', after all, means 'present at birth' and these behaviours are not. Unlike more basic physiological functions, such as the suckling reflex, or the ability to breathe, cry and excrete, they appear sequentially during development and in cultural context. Nor can one simply substitute the word 'instinct' for 'innate' without confusion, The ethologist Pat Bateson has pointed out that the term 'instinct' covers at least nine conceptually different ideas, including those of: being present at birth (or at a particular stage of development); not learned; developed in advance of use; unchanged once developed; shared by all members of the species (or the same sex and age); served by a distinct neural module; and/or arising as an evolutionary adaptation.[1] The problem with all of these – not necessarily mutually exclusive – senses of the word is that they lack explanatory power. Rather, they assume what they set out to explain, that is, an autonomous developmental sequence. They are thus as empty as Molière's famous burlesqued explanation of sleep as being caused by 'a dormative principle'.

Developmental systems theory points out that autonomy is precisely what cannot occur. Development occurs in context; the developing organism cannot be divorced from the environment in which it develops.

The acquisition of the ability to walk, for example, which occurs in most children at around a year does not hatch fully fledged from a virgin genome. Achieving the skill of walking is yet another example of the paradox of being and becoming, building as it does on several months of experimentation and practice, of crawling, standing upright and so forth, during which the toddler-to-be rehearses and develops the relevant skills, muscular control and sense of balance, normally aided and encouraged by her caregivers. And while fully developed walking is a general-purpose skill, gait learned in the context of a carpeted living-room floor will not be the same as gait learned in the savannah or desert; like all aspects of behaviour, walking is walking in context.[2] At each moment or stage of development, the autopoietic organism is building on its past history at all levels from the molecular to the behavioural. Behaviours – whether those of humans or of other animals – are enabled by the specificity of ontogenetic development, and modulated by the plasticity of physiological mechanisms in response to that past history and present context.

Arguments for the presence of innate or instinctive modules for language, facial recognition or number processing are often based on the lack of these abilities in adults with specific brain lesions, or, more positively, on evidence that these regions of the brain become active, as indicated by fMRI for instance, when the individual is set specific relevant tasks. However, this does not imply that such specificity is present at birth. Rather, the brain may begin as a general-purpose processor of information, extractor of meaning and maker of internal representations. There will be regions of the brain that are particularly relevant for these functions, which, during development and in engagement with the very act of developing in context, become progressively more specific. Thus during development certain critical or sensitive periods occur during which the brain is specifically primed to respond to environmental context and in doing so to acquire particular capacities or behaviours. During such periods particular relevant brain regions exhibit great plasticity, modifying their structure and connectivity in response to experience – for example, visual or linguistic. Synaptic connections are moulded, some being sharpened and others pruned away, so as to identify and respond to relevant features of the external world by building appropriate internal representations of that world and plans enabling action upon it. The result of this experience and activity dependence is the domain specificity of function that the mature brain reveals and which forms the theme of the next chapter.

The growing brain

Babies' brains lack their full complement of neurons, glial cells, synaptic junctions. At birth the brain weighs only about 350 grams compared with the 1300 to 1500 grams of the adult. It will be 50 per cent of its adult weight at six months, 60 per cent at one year, 75 per cent at two-and-a-half years, 90 per cent at six years and 95 per cent of its adult weight at ten years old. The rapid growth of the brain is reflected in the growth of the head. Babies' heads are large for their bodies compared to adult's. Average head circumference is 24cm at birth, 46cm at the end of the first year, 52cm at ten years and not much more in the adult. What all this means is that in the newborn the brain is closer than any other organ to its adult state of development. At birth the brain is 10 per cent of the entire body weight compared to only 2 per cent in the adult. At puberty the average brain weight is 1250 grams in girls and 1375 grams in boys, though the relationship between growth rates in the two sexes is complex and they mature at slightly different rates; thus girls are generally regarded as being developmentally and temperamentally more advanced than boys up to and around puberty. By adulthood, matched for body weight, average brain weights are similar in the two sexes.

As much of the information about the physiological and biochemical processes involved in human brain development comes from the study of laboratory animals, it is instructive to compare the overall developmental state of the human brain at birth with the developmental state of the brain of other mammals, although this varies markedly from species to species, even between those with apparently relatively similar life styles – for instance, those 'standard' laboratory animals, rats and guinea pigs. The young of the rat are born blind, naked and helpless; those of the guinea pig are born with their eyes open, they are covered in fur and able to run, following their mother. Compared with the rat, the guinea pig is clearly practically senile at birth. Mapping differences in developmental age at birth against brain maturation for a number of indices – the presence of certain enzyme systems, and so forth – is necessary if meaningful comparisons are to be made. From this it appears that in terms of brain development at birth the human falls between the pig and the rat: about 10 per cent of the spurt in brain growth is post-natal in the guinea pig, 50 per cent in the pig, 85 per cent in the human, and 100 per cent in the rat.

Because the rat's brain development at birth is limited, it is easy to follow the post-natal changes in its brain biochemistry. Over the first three post-natal weeks – that is, until weaning – a series of dramatic changes occur as neurons mature and synapses are formed. The activity of the enzymes responsible for the synthesis of lipids such as myelin increases markedly for the first fortnight or so, then levels off and ultimately declines, while the rate of protein synthesis, extremely high at birth, declines dramatically to about a third of its original level before levelling off between fourteen and twenty-one days, at a rate which does not subsequently change throughout the remaining 2 to 3 years of a rat's normal life span. The concentrations of amino acids in the brain associated with transmitter function, and of the enzyme acetyl-cholinesterase, a key marker for certain transmitter systems, also increase substantially over the period of the first twenty-one days of the rat's post-natal life, until reaching an adult level at around the end of this period. Over the first few days there is thus a massive growth spurt in brain development, implying a critical or sensitive period, during which a variety of environmental contingencies can have far-reaching effects on brain structure and performance. It is reasonable to assume that similar sequences of development occur in humans, though unlike the rat but like the pig, these sensitive growth periods are both pre- and post-natal, and may last for the first eighteen months of a young human child's life.

As the number of neurons in the human brain does not greatly increase after birth, the post-natal quadrupling of the size and weight of the brain must involve other structures. There is a proliferation of glial cells, which are still relatively infrequent at birth but whose numbers increase rapidly over the first eighteen months of infancy. Above all, however, the growth in brain size results from a change in the relationships between neurons. At birth the human brain is relatively low in myelin. Most of the pathways that have been formed are unmyelinated, and the grey matter of the cortex is rather hard to distinguish from the white matter of the subcortical tissue. The major increase in the laying down of the myelin lipid sheaths of the axons occurs during the first two years of life, a process closely coupled with the development of the glia, one at least of whose functions is to conduct this myelination.

Not only the development of myelinated axons but the vast ramification of pathways and interactions between cortical cells also occur post-natally. As the dendritic and axonal complexes spread, so the distances between the cells of the cortex increase and the surface area enlarges, forcing the development of more and more convolutions. The spread of dendritic processes brings with it an enlargement in the number of synaptic contacts between cells, which also greatly increase over this period. In addition the cells themselves change in appearance. The neurons become noticeably larger, their protein synthetic machinery develops and a number of small fibre-like structures – neurofibrils – are formed.

These patterns of growth present extraordinary problems of organisation. The various mini-organs that comprise the brain are growing at different times and at different speeds. The result is that they continually change their relationships one to another. For an example, think of the relationship between eye and brain. The neural pathways involve connections between the retina and visual cortex via the lateral geniculate. These regions grow at different rates, which means that synapses between them must be regularly broken and remade during development. Yet visual experience cannot be interrupted or continually reorganised, so the patterns of synaptic connections, the overall topography, must remain the same even whilst individual synapses are being replaced – another example, therefore, of the processes of being and becoming, the remaking of the aeroplane in mid flight, that autopoiesis implies.

The development of function

It is relatively straightforward to study the emergence of function in babies over the first few days, weeks and months of post-natal life. Parents, caregivers, teachers, do so as a matter of course; developmental and child psychologists can build on and systematise these individual and anecdotal accounts. The last decades have seen a great proliferation of such studies, plotting the growth of competencies in infants from a few post-natal minutes towards puberty. However, to match these changes with the maturation of the brain is difficult, even granted the increasing availability of relatively non-invasive windows into the brain. Babies can for instance be fitted with hairnets containing an array of recording electrodes. However, most such techniques require the subject to remain still for many minutes and it requires a degree of experimental ingenuity and patience to persuade a baby of the necessity not to wriggle.

Birth represents a dramatic change in the environment for the newborn – the relative constancy of the womb with its controlled temperature and supply of nutrients is replaced by the far less friendly outside world of varying temperature, humidity, irregular supply of food and innumerable hazards to life and limb. The newborn is being a baby, becoming a person. Reflex mechanisms unused till birth must be brought into play. Babies must breathe, feed, regulate their own temperature and, above all, begin to respond to the rich inconstancies of the natural and social world around them. Everything changes, even their internal biochemical and metabolic mechanisms. Respiratory, suckling and digestive reflexes, urination and sweating become essential. None of these is cortex-dependent. They depend on the hind and midbrain. Babies born without a cerebral cortex show them, for example. The suggestion is that the activities of the newborn infant are controlled mainly by the spinal cord and the lower part of the brainstem. The thalamus may also be involved, but the cerebral cortex at any rate plays little part in the life of the newborn child.

The baby emerges into a public world in which she is constantly in contact not merely with the dark womb and its background sounds of maternal heartbeat and bowels, and dim echoes from beyond, but with a continuous battering of new stimuli from the environment. Signals assail every one of her senses. New information floods in at a vast rate. It is in this new and rich environment that the cortex begins to grow, neuronal connections sprout, synapses are made, internal representations of that outside world are created and consciousness, in all its manifold senses, emerges. The cortex is not evenly developed at birth: those regions associated with the most urgent survival needs mature earliest, others sequentially as more and more information is processed and given meaning and the child achieves an increasing range of competences. At birth motor regions are more highly developed than the temporal and occipital cortex. Of the motor cortex, the most mature part is the region that will later control movements of the upper part of the body. This is the only area in which, by birth, the large pyramidal neurons which in adulthood carry signals from the cortex towards the muscles are mature and functionally connected. Very shortly afterwards, the main sensory area of the cortex develops, and subsequent to that, the visual area itself. In functional terms, this suggests that the most important requirements for the newborn infant are those of motor co-ordination of the top part of her body.

By the first month after birth, there is an increase in the thickness of most parts of the cerebral cortex, particularly in the motor region and the primary sensory regions, but it is not clear whether any of these neurons are yet capable of transmitting nerve impulses. For example, the axons of some of the large neurons from the motor cortex can be seen to extend as far as the spinal cord, but it is doubtful if the link is functioning yet. The myelin sheaths which have begun to develop on the upper part of the axons do not extend as far as the spinal cord, while it is far from certain if there is any communication between one area of cortex and another. The one-month-old baby is plainly still a subcortical organism.

The three-month-old child is very different. The neurons have been separated by the continuous growth of their processes and of glial cells, while this expansion has forced the cortex to increase its surface area by wrinkling up to form its characteristic convolutions. The cells of the motor region are well developed, particularly those controlling the hand, followed in turn by the trunk, arm, forearm, head and leg. All of them begin to show signs of myelination, most advanced in the region of the trunk, arm and forearm. A similar pattern can be seen in the sensory region, where the neurons responsible for the receipt of sensory information from the top part of the body and the hands are relatively well developed, while the head and leg regions are less advanced. The axonal pathways between different brain regions, and the nerves entering the cortex from lower parts of the brain have begun to develop as well. This is particularly true for the pathways arriving at the cortex from the top part of the body, and in conformity with this, the myelination of these nerves is most advanced.

The implications of these changes should be that the motor activity of the three-month-old child is in advance of her sensory processing ability, while of the variety of potential motor actions, those associated with the hand and the upper part of the body should be those which have developed most. And so indeed it proves to be. By the third month after birth the child can control the position of her body, head and legs. She can begin to grasp things with her hands; soon her fingers will be capable of making those precise and co-ordinated movements that are so supremely human. There is evidence of the increasing control by the cortex over the activities of lower brain regions, as the more basic reflexes, such as the grasp reflex, which characterise the one-month-old child, disappear, presumably as a result of inhibition of the functioning of lower motor centres by the developing cortex.

Over succeeding months the dendrites and axons of neurons in all parts of the cortex continue to increase in length, in thickness, in degree of myelination. Glial cell development continues. The sequences of all these changes continue to be such that the motor cortex and the primary sensory cortex are more advanced than other cortical regions, but by six months other areas have begun to catch up. More and more bodily activities come under cortical control. By six months the average child born and reared in Europe or America can spread out her fingers or touch her fingertips with her thumbs, although these movements are not yet fully developed. She can roll her body over, and sit up without support. By the seventh month she can crawl, by the eighth pick up small objects, by the ninth, stand, by the tenth month walk with help, and by the twelfth, walk without support. However, none of these skills occurs except within specific contexts. They are in some measure culture-bound. The development of particular skills at particular times is expected in our own, and every, culture; however *which* skills are expected *when* will vary. At different historical periods and in different cultures, children may be expected to walk at different developmental stages. Their performance, within limits, will conform to these expectations. An example of the variation in performance between different groups was pointed out many years ago by M Geber and RFA Dean,[3] who observed that babies born into certain traditional African cultures are from four to eight weeks more advanced in beginning to sit and walk, than are the norms in industrial Europe or the US. White babies are regarded as 'lazy' by contrast. According to the developmental neuroscientist Lise Eliot's book on child development, *Early Intelligence,* similar precocity has been reported from pre-industrial societies in Latin America and India.[4]

All this demonstrates that as soon as one moves beyond the area of simple sensori-motor responses it becomes harder to analyse development in terms of 'species-typical' responses. Even the patterns described so far vary with the individual child and her circumstances, which may make speech advanced in some, walking delayed in others, and so forth. Considerable deviations can occur between children, for instance, in the age they begin to walk or speak, without any apparent later differences in performance.* Thus even the simplest of generalisations may be open to qualifications. The older the child, the greater the number of individual developmental differences that can no longer be clearly associated with specific gross changes in brain structure. What is more, although it is possible to say a fair amount about the overall pattern of brain development during pre-natal and just-post-natal life, if only by extrapolation from the studies with animals, little is known about the details of the more subtle changes that occur later in childhood up to puberty or adulthood as the brain continues to respond to internal and external environmental changes – for example surges in hormone production, changes in the immune system and so on, as well of course as social context. Inevitably, the experience of the individual is superimposed upon generalisations covering a population of individuals – a feature of the plasticity, rather than the specificity of the brain. Yet it is just these questions, about the brain processes associated with the development of self and social awareness, of consciousness, of agency, that it is so important to try to understand.

Turning sense-inputs into perception

For many reasons neuroscience has paid more attention to the visual system than to any other of our senses. Perhaps this is because vision is so important to us. Also, by comparison with the subtleties of olfaction and taste, or even hearing, visual experience is the most readily experimentally manipulated both, more dramatically, in animals, and in

---

* I was very impatient for my younger son to begin to speak; my mother-in-law was far more relaxed, assuring me that by the time he was four he would be chattering so incessantly that I would look back on the period of silence as almost halcyon. She was right – and he hasn't got any less articulate in the subsequent thirty-six years!

psychophysiological experiments in humans fitted, for instance, with distorting lenses. This may be why the relationship between vision and perception, between seeing and interpreting images, has become a paradigm for neuroscientists interested in questions of conscious experience. Francis Crick, for instance, in his book *The Astonishing Hypothesis*[5], argues quite explicitly that it is possible to reduce questions of consciousness to those of awareness, and of awareness to perception, so that if we can understand the neural processes underlying vision and perception, we will have a model for consciousness. I'll come back to this in more detail in the next chapter, but for now I want to concentrate on what is known about the development of vision and its relationship to perception.

Humans, unlike rats or cats, are born with their eyes open, seemingly instantly to observe the world. All parents will be aware of the way that their newborn stares gravely at them, seeming infinitely wise, both curious and judgemental. Yet just what newborns can observe and interpret is strictly limited. The retinae and their optic nerves, connecting them via the lateral geniculate staging post to the 100 million neurons of the primary visual cortex, are fully present at birth, though both continue to grow over the following months and years. Synaptic density only reaches its peak in the visual cortex by about eight months. Over the same period the nerves connecting retina and lateral geniculate, and visual cortex steadily increase in myelination. Beyond eight months the steady pruning of seemingly redundant synapses in the primary visual cortex begins to take its toll, reducing them by around 40 per cent as the remaining connections proliferate and stabilise.

These developmental changes in the cortex can be matched to changes in visual ability. At birth babies are better at motion detection than at interpreting form, and can focus only on objects within about 20 to 100cm in front of them. Within this range they attend most closely to strong lines and shapes, and, importantly, to faces. Just after birth, babies gaze preferentially at head-like shapes that have the eyes and mouth in the right places. If the images are jumbled or the eyes and mouth reversed, newborn babies respond much less strongly to them.[6] Newborns cannot distinguish colours; this capacity only develops by about three months, as both the specialised wavelength-sensitive cone cells of the retina and the corresponding regions of the visual cortex develop. Furthermore, because the retina develops unevenly at birth, babies see best at the periphery; it takes another couple of months for retina and visual cortex to mature sufficiently for more central vision, visual acuity and proper depth perception (binocularity) to develop.[7] This makes possible more precise distance assessment and hence a greater degree of visuomotor co-ordination. Thus at three months, a child can focus her eyes on a more distant object and move her hands towards it.

The auditory regions are already well developed at birth and thus able to process information associated with sound arriving at the ears from the outside world and to interpret it – that is, to enable the baby to distinguish between different voices, to translate inputs into perceptions, and to respond accordingly: for example, as every parent will hope, by stopping crying if all is well. In fact, from the early days of life infants are sensitive to sound features such as pitch and timing differences that are fundamental to music and underlie the intense human interest in music that crosses cultures.[8] Ears, like eyes, auditory perception, like visual perception, are developmentally primed.

Perception is thus experience- and action-dependent. Deprived of adequate visual stimulation the wiring of the visual cortex fails to develop adequately. No one would wish such an experience on a developing human, so once again animal experiments have to point the way. If rats are reared in the dark, the numbers of synapses in their visual cortex is sharply reduced (Fig. 5.1). Rearing kittens in an environment in which they see only vertical or horizontal striped patterns during the critical period of synaptogenesis affects the pattern of connectivity of synapses in particular regions of the visual cortex such that the neurons will learn only to respond to vertical or horizontal stripes. Rearing kittens with one eye closed results in a profound rewiring of the visual cortex such that the deprived eye loses its connectivity to the cortex, and the resultant space is taken over by inputs from the non-deprived eye.[9] One way – the one conventionally deployed – of interpreting such plasticity is of a competition for synaptic and neuronal space between rival or alternative inputs. The alternative is to see the brain self-organising so as to make the best use of its available synaptic space by favouring experience-relevant inputs. In either interpretation, during these critical periods the visual system is responding adaptively to the world in which the developing organism finds itself.

Such findings can be extrapolated to humans. Babies born with non-

Fig. 5.1 *Dendrites from the visual cortex of rats: left, from rat reared in the dark; right, from its littermate a week after being exposed to the light for the first time. Note the proliferation of spines – the junction points for synapses. From an experiment by F Valverde.*

aligned eyes (squints or wall-eyed) do not develop adequate binocular vision, because the visual cortex wires up appropriately to the conflicting inputs from the two; the failure is lasting if not corrected by adjusting the alignment of the eyes during the first six to eight months. As experiments on visual deprivation in cats has been a major point of attack by animal rights activists, it is not unimportant to note that the operations required to correct squints in babies have drawn heavily on such experiments.*

What all this implies is that in humans, just as in other mammals, there are sensitive or critical periods during which the visual regions of the brain are particularly responsive to environmental contingencies,

---

* Not that this would justify such experiments in the opinion of very hard-line animal rights activists, but for the rest of us – especially those with children who have benefited from such operations – it is surely relevant.

learning, and hence developing appropriate wiring patterns, to extract and interpret meaningful regularities from the environment. The slow maturation of the human brain by contrast with that of even our closer relatives means that this period may be considerably extended. Furthermore, the different rates at which different visual feature detectors develop may imply not one but a number of distinct critical periods, for shape, form, colour and so forth. Face recognition occurs very early, it seems almost within a few hours of birth; judging by their reactions to videos, even day-old babies can apparently distinguish between their mother's and another woman's face.[10] Presumably what is important here is not some mysterious property of being a mother, but that hers is the face the newborn would normally first experience for any significant length of time. This is yet another indication of the central importance of vision amongst the human senses. The phenomenon would seem not dissimilar to that of imprinting in young chicks, who within the first two post-hatch days become attached to the first prominent moving object they see – normally the mother hen of course, except in modern farming's grotesque battery-rearing conditions. By contrast, in other species such as sheep, it is olfaction rather than vision that is important to such early attachment, although sheep too can recognise faces – the faces of other sheep, that is.[11] Olfactory recognition depends on pheromones – specific secreted chemical signals. Thus newborn rabbit pups respond to a pheromone emitted by their mothers' mammary glands that is attractive and induces them to grasp the teat with their mouths.[12] That such pheromone signalling exists in humans too, and contributes to mother-infant attachment, is probable, as babies can discriminate, and prefer, their mother's smell to that of a stranger, though this is much less well explored.

Face recognition in babies, however, shows some interesting maturational effects. The ability to perceive faces narrows during development, due to cortical specialisation that follows from experience of seeing faces, and learning to extract specific features – shape of nose, eye, mouth, hair colour, etc. – by which to recognise them. Expertise in identifying faces requires practice built up over several years. During early development, it would seem that a generalised ability to recognise and distinguish faces gradually becomes more specialised and selective; six-month-old infants recognise human and non-human primate faces indifferently; by the time they are nine months old they are much better at recognising human than non-human faces.[13] This is thus yet another example of how what appears to be an innate or instinctive 'modular' brain or mental capacity is actually dependent on experience during development. There was a frisson of racial concern a couple of years back when Alexandra Golby and her colleagues reported that adults shown 'same-race' (read skin colour) faces had different brain responses in a specific region of the cortex, as measured by fMRI, to when they were shown 'other-race' (ie other colour) faces.[14] The point however is familiarity – because of the way class and occupation are still colour-coded in the US, whites are less familiar with black faces, whereas members of the minority black community need to be more familiar with whites; correspondingly the differences in brain signals between blacks shown black or white faces is less than it is for whites. The region activated in Golby's experiment was a region in the cortex (the fusiform cortex), but this is not a 'face-specific' area, as according to Golby it is apparently also activated when bird or car experts are shown bird or car pictures!

Thus the brain is primed, so to speak, to learn to extract significance from sensory input. To do so, it has to bind together, to associate, many different forms of such input. Considering vision alone, it is necessary to be able to link motion to form so as to perceive moving objects, to associate colour with that form, and to assess the distance and direction of the coloured moving object. Just how this binding is achieved – one of the central areas both of mystery and research in neuroscience today – is a topic not for this but for the following chapter. Extracting significance means recognising invariance (this shape and feel, say, of a toy, a teddy bear, is always associated with this colour, yellow). But it also means recognising variance, the sudden appearance of a novel object within the field of view, a moving rattle above the baby's cradle, for instance, or the looming face of a caregiver. Furthermore, visual inputs have to be linked to all the other associated sensory inputs: the sound of a caregiver's voice, the taste of milk, the feel of skin – a computational problem of enormous complexity yet seemingly achieved effortlessly by the developing baby.

Turning vision into perception demands another capacity, central to my own research interests, and which I have hitherto studiously attempted to avoid discussing: that of memory. To be perceived, incoming visual signals need to be retained for a relatively brief period and associated with others either temporally contiguous or derived from past experience. Thus they have to be held in what is termed working memory. So when and how does memory develop? Some memories are more important than others; clearly faces are amongst them as newborns so soon learn their caregiver's features, but few features of the infant's environment are so important. The more general ability to recognise and in consequence remember episodes in one's own life, so-called autobiographical or episodic memory, develops steadily, in parallel with the maturation of the neocortex. At six months, infants can remember events for up to twenty-four hours, which extends to up to a month when they are nine months old. A region central to the transition between short- and longer-term memory, the hippocampus, shows significant increases in dendritic and synaptic number towards the end of the first year. Conor Liston and Jerome Kagan tested the ability of infants to recall motor acts first experienced at 9, 17 or 24 months of age. Thirteen-month-old children failed to recall events experienced at 9 months, whilst 21- and 28-month-olds could remember events witnessed at 17 and 24 months.[15] Children at this age can clearly remember places they have visited and experiences – particular food eaten or games played – in those places.*

The existence of such critical periods means that early learning shapes later perception. Think of the difference between a rural and an urban childhood, for instance. The rural child (or, let me not romanticise, at least a child reared in less industrialised, more traditional rural areas) extracts such important information from her perceptual world as the differences between ash and beech trees or the calls of chicken and ducks. Over the same developmental period the urban child may learn to recognise the difference between a Mercedes and a Ford, the alarm-sounds of police cars and ambulances. Different early experiences shape the attention we pay to different but not greatly dissimilar features of our environment, and embed themselves within our memory in ways that are more deeply engraved than the memories of adulthood.† Our

---

* I've been watching with grandparental delight the capacity of our own youngest granddaughter, currently aged four, impeccably remembering the location of toys and special biscuits kept for her in our own house, which she visits at irregular intervals.
† I've written about early childhood memory and its eidetic quality, by comparison with the linear memory of adults, in my book *The Making of Memory*, and I won't be discussing its implications in detail here.

youth provides the contours within which our perceptual worlds are formed, and indeed this makes good survival sense in evolutionary perspective. The extraordinary versatility of humans has made it possible for them to live in environments as different as the Sahara and Greenland. Yet over most of the evolutionary history of our species – indeed until the eye-blink of the last couple of centuries – a child born in one type of environment would be unlikely ever to move to another very different one. Early extraction of the regularities of that environment, moulding synaptic connectivity and framing perception, would help ensure survival.

The emergence of language

The transformation of vision into perception is a process in which the ontogenetic specificity of the growth of connections between retina and visual cortex, and within the many sub-regions of the cortex, occurs in intimate interplay with experience-dependent plasticity and is relatively uncontroversial amongst neuroscientists. This is far from the case with language, however, perhaps because its study lies at the intersection of many different research traditions, from philosophy via semiotics and linguistics, cognitive and evolutionary psychology to neurology and even genetics. Current assumptions are that in adults, Broca's area is necessary for the production of syntax, grammatical formations involving verbs (and thus the habitus of Chomskyian universal grammar, should it exist), whilst Wernicke's area is relevant for semantics, the meaning of words, especially nouns. These regions are slightly larger in left than right hemispheres in most people and the matching areas on the right are supposed to be concerned with prosody and speech rhythms. In some people left and right specialisations may be reversed, and also brain plasticity in early childhood means that if the left hemisphere regions are damaged there can be some takeover of function by the right. During post-natal brain development, synapse production is at its height in Wernicke's area between eight and twenty months, whilst Broca's area matures more slowly, not being fully developed until around four years; myelination is also slower in Broca's than Wernicke's area. It makes sense to assume that these patterns of maturation are related to the development of spoken language and its comprehension.

The patterns of emergence of speech in babies brought up in an interactive environment with their caregivers are relatively predictable. From as early as two months of age, babies begin to babble – to practise sounds, initially vowels, and later, by about five months, consonants involving the use of tongue and lips.[16] Between one and two years, babbling develops into strings of syllables in which recognisable words are interspersed and, by eighteen months onwards grammatically meaningful sentences. From then on words and concept acquisition proceed at an astonishing rate. By the age of two most children can match the years of training Savage-Rumbaugh gave to Kanzi, and from then on the acceleration in linguistic skills is dramatic.*

The debate over whether language is innate or acquired goes back to ancient philosophers and has engendered much acrimony, at least in part, I would argue, because the question is wrongly posed. Was there, they asked, a natural, primeval language? Would children brought up in isolation speak at all, and if so what language would they use? As with vision, the development of the brain and vocalising structures involved is an example of ontogenetic specificity; how these organs are employed is clearly shaped by the linguistic context. These early years – perhaps the first four – form a sensitive period for the acquisition of language, as exemplified in the use of the term 'mother tongue' to refer to Chinese, Swahili, Urdu or English, or indeed any other of the many thousand different native languages and dialects in our intensely linguistic universe. Children raised in bilingual households from birth acquire both languages; but learning a second language later in life is, as many of us know to our cost, remarkably hard, and in fact seems to engage different brain areas from Broca's.†

Just as apparently our brains become tuned to recognising the features of more-regularly experienced faces, so too with linguistic ability. Over

---

* A couple of days before writing these sentences I was at a dinner table with my young granddaughter, who carefully informed me that she was sitting between me and her mother. Who was sitting *opposite* her? I asked. She didn't know the word, so I explained. Within seconds she was employing it to identify who was sitting opposite whom amongst the twelve people at table, and generalising its use to wholly different contexts.

† I am writing this in France, my auditory landscape filled with a Gascon-accented French to which I fear neither my ears nor tongue will ever readily accommodate; by contrast, Italian, which I acquired in my youth, comes much more readily – I daren't write *naturally* – to me.

our early years we become accustomed to the vocabulary and sentence construction of our own mother tongue, or, in the case of bilinguality, 'mother tongues'. But more than that, we learn the prosody, the speech rhythms, typical of our language and dialect. Our auditory cortex and right hemisphere regions become accustomed to them and it becomes harder not merely to speak but even to hear a foreign tongue.[17] Anglophone children find it increasingly hard to distinguish the sounds in a tonal language like Chinese, whilst Japanese children lose the ability to hear the distinction between 'l' and 'r' in English. So-called 'wild children', or those so brutalised by their parents or caregivers as only to emerge into normal human society when they are beyond such sensitive periods, have great difficulties learning to speak at all. Children born deaf go through the early phases of babbling as do hearing children, but, unable to receive spoken feedback, turn instead to 'babbling' with their hands, a precursor to the development of grammatically structured sign language. Early experience both sharpens and limits our 'mental' capacities.

So much is generally accepted, but at this point the controversies begin. It is convenient to date them from the demolition of the prior, behaviourist theory of BF Skinner by Noam Chomsky in the late 1950s. Skinner had proposed that children learn appropriate words and grammatical constructions by contingencies of reinforcement – essentially by being rewarded for correct use and punished or denied reward for incorrect use. This is of course the way that chimpanzee-language teachers go about their task. By contrast, as I hinted in the previous chapter, Chomsky argued that there was a universal grammar by which children recognise nouns and verbs, subject and object and learn to place them in correctly meaningful order. Thus the normal sequence in English is subject-noun, verb, object-noun, plus appropriate qualifiers, and any sentence in this form, even when pairing incongruous nouns and verbs (Chomsky's often quoted example being 'colorless green ideas sleep furiously'), will seem to make sense. When later Fodor introduced his claim for the existence of mental modules, Chomskyian universal grammar came to be seen as the product of a mental 'language module' to be trumped in due course by Chomsky's sometime student Steven Pinker's claim for an evolved LAD[18] (As I pointed out in the previous chapter, Chomsky has until recently seemed rather indifferent as to how his universal grammar might be represented in the brain.)

Pinker's argument is based on Chomskyan universals, an innatist view of how language develops, some fragmentary neurological and genetic evidence, and evolutionary speculations about linguistic origins. Thus the linguistic difficulties experienced in a family with the FOXP2 mutation, also referred to in the previous chapter, were hailed as demonstrating the presence of 'language genes',[19] despite their presence in many non-human and non-linguistic species. In fact, human individuals carrying this mutation find it difficult to select and sequence the facial movements necessary for articulation and have problems comprehending phonemes and grammatical structures (which suggests that their problem cannot be simply reduced to a putative missing 'language gene').[20]

A major controversy has emerged over the rare condition known as Williams syndrome, associated with a chromosomal abnormality – a small piece of chromosome 7 is missing. Adults with the syndrome show good levels of proficiency in face processing, language and social interaction, despite severe impairments in spatial cognition, number and problem solving This relative sparing of specific faculties is, it is argued, evidence for a discrete 'language module' in the brain. The developmental psychologist Annette Karmiloff-Smith, in disputing this interpretation, quotes examples of such claims:[21] 'It is uncontroversial that the development [of Universal Grammar] is essentially guided by a biological, genetically determined program'[22] and '[the human mind is] . . . equipped with a body of genetically determined information specific to Universal Grammar'.[23] Or, as Pinker puts it in *The Language Instinct*: 'The mind is likely to contain blueprints for grammatical rules . . . and a special set of genes that help wire it in place.'

For such claims, Williams syndrome would seem to be the acid test. If the sparing is present from birth it would argue for an innate and specific language module – Pinker's LAD. Yet meticulous studies by Karmiloff-Smith and her colleagues reveal that far from linguistic and face-processing abilities being straightforwardly spared in Williams syndrome, children with the syndrome are indeed impaired in quite subtle ways and, as a result, as they grow up develop alternative strategies for overcoming their deficits, thus recovering some functions and, as adults giving the appearance of possessing a specific 'language module'. As she concludes:

subtle differences in developmental timing, numbers of neurons and their connections, transmitter types and neural efficiency are likely to be the causes of dissociations in developmental outcomes. Such subtle initial differences may have a huge but very indirect impact on the resulting phenotype, ultimately giving rise to domain-specific impairments after the process of post-natal brain development.

Where does this leave innate universal grammar and language instincts? I would argue that the claim is analogous to suggesting that we humans possess a 'perception instinct' – that is, it is a category error. There is an ontogenetically specific capacity to acquire language, a capacity as dependent on the genes as binocular vision or any other aspect of human behaviour. The acquisition under normal conditions occurs seamlessly and sequentially during development, passing through critical periods of linguistic and auditory practice during which sounds, made by oneself or others, are gradually understood to symbolise objects and intentions. Unlike vision, language is a specifically human capacity, a special case of symbolic communication resulting from an exaptation made possible by the particularity of our larynx and vocal apparatus. In the absence of the ability to hear and respond to sound, other forms of symbolic communication are possible, as when deaf children use hand-signing.

It is communication via symbolic manipulation that is essential to social organisation in the human species. Language has developed to facilitate this process, and in turn has been facilitated by it. Why then the seemingly universal grammar? In their several ways both Deacon in *The Symbolic Species* and Donald in *A Mind so Rare* provide the answers. Language and the brain have co-evolved. Language has evolved in order to fit brain processes that partition out the world in particular ways, between things and actions, nouns and verbs; and as language has evolved so it in turn has shaped the evolution of brain (and mental) structures to fit its increasing demands on brain and mental space and brain and mental processes. This is why today we find it almost impossible to conceive of thought without language. 'I've got to use words when I speak to you' as the lyric puts it. Whether there ever was an environment of evolutionary adaptation some time back in the Pleistocene, as evolutionary psychologists would have us believe, the capacity to symbolise, to vocalise and communicate the meaning of symbols, emerged over that period, and in doing so helped also to shape not just adult humanhood and human society but our very ontogeny, building on and reinforcing the existence of activity and experience-dependent synaptic modulation in key brain areas.

Consider, for example, a vital aspect of becoming competent in industrial society: learning to read. This is obviously a culturally specific skill, and one clearly not demanded in any putative Pleistocene EEA. It would be hard to argue that humans possessed an innate 'reading module' or 'reading instinct'. Children need to be taught to read and in our society typically, if they are working in languages using the Roman alphabet, begin acquiring reading skills from the age of about two to three years onwards, initially recognising the shapes of letters and words, and associating them with sounds, and then associating them with the concepts or objects they symbolise, later being able to read and understand entire sentences, with nouns, verbs and all other sentence components in place. Mature reading engages a left hemisphere network of frontal, temporal, occipital and parietal regions that serve to map visual on to auditory and semantic representations. As children learn to read changes occur in their brain activity, with increases in left hemisphere and decreases in right hemisphere activity, as measured by fMRI.[24] Reading skills using pictogram alphabets such as Chinese or Japanese may engage slightly different brain regions. Once again, therefore, the acquisition of a skill itself both depends on an appropriate degree of maturation of the brain – a critical or sensitive period – and results in lasting experience-dependent changes in the brain activity required to sustain it.

Communication and social interaction

Which brings me to the final, and perhaps most fundamental aspect of becoming human. As I have emphasised, human infants by comparison with all other species require long periods of care before they achieve autonomous status, and it is in the process of this nurture that they become socialised, learn how to interact with others, and develop that so-called theory of mind which enables them to understand that other humans also have intentions, needs, agency, and are not mere objects. Interaction begins at birth, as the newborn baby is put to the mother's breast. The mutual holding between infant and caregivers*, the cooing speech with which the baby is addressed and to which she begins to respond (sometimes called 'motherese,' almost a duet between mother and child), the visual and possibly pheromonal interactions by which she so quickly learns to distinguish her caregiver's face from others, all occurring within the first few hours, days and weeks of birth, are crucial to subsequent development.

Very early the child begins to imitate facial gestures and sounds made by its caregivers, and to provide one of the most potent of rewards – smiling. All babies, even those who have been born blind, and consequently never able to see a human face, nevertheless start to smile at around five weeks, perhaps enabled by the continuing myelination of crucial brain structures such as the basal ganglia (a group of neuronal masses within the cerebral hemisphere, including the amygdala – see Fig. 3.1). Furthermore, at this age they do not have to see other people smile in order to smile themselves; this, like the development of binocularity or babbling, seems an ontogenetically specified process, which builds on the interest in faces that babies have and the facial grimaces – proto-smiles – that they show even within a few hours of birth. But smiling soon changes from an internal to an exogenously produced response; by about eight weeks it has definitely become a social act. Babies smile in response to other people smiling at or playing with them and, as they grow older, sighted children also learn to modify their smiles according to their experience, producing, as Bateson puts it 'subtly different smiles that are characteristic of their particular culture. Nuance becomes important.'[25] A blind child, lacking visual interaction with her mother, becomes less responsive and less varied in her facial expression.

---

* I have deliberately mostly used the more neutral term 'caregiver' throughout this chapter for a complex of reasons. It isn't just about the political correctness of recognising that caregiving is not gender specific, and fathers as well as mothers can be caregivers. It is that, as Hilary Rose has pointed out, the very concept of a mother has been rendered complex by the arrival of the new reproductive and genetic technologies, which have severed the links between the provider of the gametes, the birth mother who carries the fertilised ovum, and other surrogates. A newborn may be ontogenetically ready to bond with her earliest caregivers, and indeed they with her, but a genetic link between the two is far from being a prerequisite – as the widespread practice of wetnursing amongst wealthy Europeans during the eighteenth and nineteenth centuries should have made obvious long before the emergence of the new technologies.

Smiling is thus part of the ongoing development of communicative and social skills that are an essential part of becoming human. It is just these skills in communication that are lacking in autistic children, who, it is argued, lack a theory of mind, and are unable to ascribe agency and intentionality to others.[26] Autism has become a syndrome of intense interest. It, and its weaker version, Asperger's syndrome, are increasingly diagnosed, more commonly amongst boys than girls, though whether this is because it is actually becoming more common or is merely better recognised, or a currently fashionable label as part of the medicalisation of the human condition that forms a major theme of the second part of this book, remains unclear. Like language acquisition, its explanation has become a happy hunting ground for those wishing to propose genetic causes and a 'theory of mind module'. Autistic children show differences in face recognition, detectable by MEG imaging. Whatever its genetic origins, autism is essentially a developmental defect, affecting a child's ability to communicate, resulting in turn in a failure in interaction between child and caregiver. In such interactions with autistic children, Peter Hobson has observed,[27] child and caregiver look less frequently at one another and fail to achieve the 'dueting' verbal communication that seems a more characteristic feature of childrearing practice, at least in Europe and the US, although his view of the significance of this in the development of autism has not gone unchallenged. Hobson argues that such interaction is indeed 'the cradle of thought' central for the growing child to develop a sense of self, consciousness and a theory of mind.

To return to the opening paragraphs of the previous chapter and the major theme of this book, it is in this sense that minds are enabled but not reducible to brains; they are the products of the open biosocial system that is the developing, communicative, symbol-using child. Minds, as Donald puts it, are 'hybrid', simultaneously product and process of this evolved and developing biosocial system. And it is within this framework that I can now begin to consider the workings of the fully-formed human brain, the task of the following chapter.

CHAPTER 6

Having a Brain, Being a Mind

ANIMALS, WE ARE INCLINED TO SAY, BEHAVE. HUMANS HAVE MINDS, intentionality, agency, personality. Animals have emotions, humans have feelings. Back in the seventeenth century, René Descartes was clear about this: animals were mere mechanisms; if they squealed when injured this was no more than the squeak of a rusty machine. Humans and humans alone had the power to think and therefore to be – *cogito ergo sum*. The entire thrust of my previous four chapters has been to contest this claim to human exceptionalism. To understand what it means to be human, we have no choice but to compare ourselves, our own brains and insofar as possible minds, with those of other species. Yet the refusal to grant animals mentation remains, powerfully bolstered by the philosophy and methods of the post-Cartesian science developed within the Judaeo-Christian tradition, which gives biblical authority to 'Man' to name, dominate and exploit all other living forms on earth. However, it is not just besotted pet-owners who, attributing personality, thoughts and intentionality to dogs or cats, would dispute that such a categorical separation exists. Although Darwin's co-creator of evolutionary theory, Alfred Russel Wallace, shared the biblical view of the distinction between humans and all other living forms, Darwin himself certainly did not, seeing the human-ness of humans as an evolved property.

Today, no serious biologist could dispute such evolutionary continuities. They – we – are left therefore with two options. We can deny that humans have minds except as epiphenomena, and argue that all the serious business of life goes on within brain and body. Minds, consciouness, subjectivity, are thus irrelevant spin-offs from having large brains. There is a long radical tradition in favour of this position. Mind is to brain as the whistle to the steam train, claimed Thomas Huxley, Darwin's nineteenth-century 'bulldog' in defence of evolutionary theory. Or, as the 'mechanical materialists' amongst the physiologists of that century put it, the brain secretes thought as the kidney secretes urine.

In a famous – almost notorious – paper Stephen Jay Gould and Richard Lewontin once described such spin-offs as 'spandrels' – inevitable but structurally unnecessary consequences of particular archi-tectural forms, such as the ribs which are required to support cathedral domes. The spandrels can then be used for other valid purposes, such as the frescoes and mosaics that decorate them in churches, but they remain structurally irrelevant.[1] Although Gould and Lewontin used the spandrel metaphor in quite other evolutionary contexts, it serves well to encapsulate the views of many present-day cognitive psychologists and philosophers. Patricia Churchland is perhaps the strongest modern advocate for such a position, with her partner Paul Churchland, airily dismissing the relevance of mental categories as 'folk psychology'.[2] For them, mind is a spandrel, available to be filled with marvellous decora-tive fantasies, but irrelevant to the real business of living and reproducing, which is the concern of the brain. If we believe that we have some sort of free will, or conscious choice over our actions independently of brain processes we are mistaken. Some go further, claiming that brains might just as well be constructed out of silicon chips as from evolved carbon chemistry. 'You don't need brains to be brainy' as the philosopher-psychologist Margaret Boden once put it,[3] arguing for computer consciousness, a theme taken to its current limits by the theorist and engineer Igor Aleksander.[4] But I'll return to the implications of this view later in the book.

The alternative is to side with the pet-owners and attribute at least some of the human properties of mentation and consciousness to non-human animals. Consciousness and mentation are then evolved prop-erties, arrived at through the processes of natural selection. Far from being spandrels, they have evolved because of their survival value for those who possess them; they enhance fitness. Such an approach, consid-ered taboo for earlier generations of psychologists through much of the past century, is now once more legitimate. Non-human animals, it is conceded, show intelligence, can think and have minds of their own – not just Kanzi and his fellow bonobos, but many other species too. The neuroscientist Marc Hauser was lauded rather than excommunicated for entitling a recent book *Wild Minds: What Animals Really Think*.[5] The same once heretical view colours Antonio Damasio's approach to the origins of consciousness and the concept of self,[6] and frames my own thinking.

Up until this point in the book I have refrained from providing a formal account of brain structures, mechanisms and processes except, so to say, in passing. To develop my argument about mind and brain, it is time to rectify this lack. Brains and nervous systems, human and non-human, can be considered as forming a nested hierarchy of structures, built of atoms, of molecules, macromolecules, cells, ensembles of cells. 'Built' is an appropriate word, for it will become apparent in the course of this chapter that it is not merely composition but structure, architecture, that is central to understanding the brain, though absolutely not, as I will shortly make clear, in the Pinkerian sense of the 'architecture of the mind'. At what level, if anywhere, within this hierarchy should one look for the biological substrate of thought and agency? The answer I will give will be as paradoxical as the brain itself: at all levels and at no level. This chapter explores the paradox, considering brain and mind in action in four crucial areas: vision/perception, pain, emotion/feeling and memory.*

The molecular biology of mind?

Until the advent of molecular biology, neuroscientists, like other biologists, regarded cells as the basic 'units' from which brains, like all other body tissues, are constructed. Neurochemistry, the biochemistry of the brain – the discipline in which I was raised – was regarded as at best a

---

* It is interesting that in two of these four the terminology exists to distinguish 'brain language' from 'mind language' whilst in the others the terms are identical whether talking the language of physiology or psychology, much as in English a distinction is made between bull and beef, pig and pork, sheep and mutton, the former terms deriving from the Anglo-Saxons who tended the animals, the latter from the Norman conquerors who ate the products of Anglo-Saxon labour.

handmaiden to pharmacology and physiology. No longer. Molecular biology is in the ascendant. Especially following the sequencing of the human genome, the claims that it is at the level of genes and their protein products that the brain will also be 'solved' have been taken more seriously. A discipline called molecular neurobiology, utilising genetic and 'proteomic' methods, offers to provide a catalogue of all the proteins present in the brain, almost as if this would indeed provide the understanding required to solve the mystery of mind. Certainly, 25,000, the current estimate of the number of genes the human genome is now believed to contain, or even the previous estimates of around 100,000, is not enough to specify even the full complement of brain proteins, let alone cells and interconnections. Combinatorial mechanisms must be brought into play. As I have already mentioned, the brain expresses a greater range of proteins than any other body tissue, some only transiently, in specific locations and in such minute quantities that even today's sophisticated techniques of analysis can scarcely identify them, and a proteomic catalogue could prove a handy tool. But, brain proteins are not a stable population. They are constantly being broken down and resynthesised even in the mature brain; indeed the brain has one of the highest rates of protein synthesis in the entire body. The average half-life of a protein molecule in the brain is around fourteen days, but many turn over much faster, in a matter of a few hours. Listing components doesn't necessarily help understand their interactions. Those of us reared as classical biochemists tend to regard proteomics as the molecular biologist's way of rediscovering biochemistry – except that as biochemists we have always been concerned with dynamics, the metabolic interactions between the myriad proteins that stabilise the cell's metabolic web in what the biochemical geneticist Henry Kacser once referred to as 'molecular democracy'. The 'mol biols', by contrast, still study snapshots, all the proteins in the cellular bag at any one moment, not their changes and interactions over time.

Proteomic enthusiasms are tempered further by a recognition that, as it isn't just the proteins but their addresses that are important, a molecular anatomy that takes into account where the proteins are located within and between cells is required. The unique architecture of neurons becomes of central importance. Thus, most of the normal cellular metabolic 'housekeeping' goes on within the neuronal cell body (and in the glial cells) and engages the same sets of enzymes and subcellular structures as in any other body cell – for instance, protein synthetic machinery involving nuclear DNA and RNA, post-synthetic processing of the proteins which takes place in membraneous structures known for their discoverer as Golgi bodies, and the energy-generating systems in the mitochondria. But the cell body needs to be able to communicate with axons, dendrites and synapses. A two-way flow of both materials and information is needed. Synapses need to signal to the cell body their needs for particular proteins, and substances synthesised centrally need to be transported to where they are required. Both matériel and information flow along the tramlines offered by the microtubules and neuro-filaments that run from the cell body along axons and dendrites. Time-lapse movies made of neurons growing in tissue culture show that the internal constituents of neurons – mitochondria, the various neuro-filaments, microtubules, synaptic vesicles – are in constant motion. Vesicles and other granules synthesised and assembled in the cell bodies, move down the axons and dendrites to pre- and post-synaptic sites along the tramlines provided by the microtubules at rates of up to 40cm a day. This may not sound very fast, but it means that amongst the interneurons of the cortex, proteins or even organelles assembled in the neuronal cell body can easily arrive at the axon terminals within thirty seconds or so. Other particles, and signalling molecules and ions, are transported in the opposite direction – traffic flow is busy enough to far exceed London's infamous M25 on a particularly bad day. Even so, however, the distances from cell body to its extremities are sufficiently great that it seems to have become important to permit some local dendritic and synaptic biochemical autonomy; it is now generally accepted that there is local protein synthesis occurring in the synapse, dependent, it is true, on RNA transported from the cell body, but none the less semi-independently from what goes on elsewhere in the neuron.

So might mentation be embedded in the brain's biochemistry? For sure, thought requires metabolic energy and thus depends on chemistry. Deprived of oxygen or glucose, the brain shuts down and cells quickly die. Less drastic tinkering, with drugs for instance, results in changes in cognition, perception and feeling, experiences that are likely to be familiar to everyone. Back in the 1960s, there were even serious suggestions to the effect that such central properties of brain and mind as memory might be stored in the form of unique brain proteins or nucleic acids.[7] More recently, microtubules, largely composed of a single protein, tubulin, together with a few minor add-ons, have been invoked by the mathematician Roger Penrose as sites of quantum indeterminacy that in some inexplicable way generate consciousness.[8] However, as identical structures exist in most other body cells and even in such distinctly un-'conscious' organisms as unicells, it is hard to see why anyone should take such a proposition seriously, unless suffering from an excessive dose of physics-deference. Subcellular biochemistry and molecular biology are not the right places to be looking for the seat of the soul. They are necessary, for they enable brain and mental activity, but they do not contain or determine our patterns of thought or sense of agency.

Neurons and their synapses

So let's move up a level in the hierarchy and build on the description of neurons I began back in Chapter 2. There are up to 100 billion neurons in the adult human brain. Perhaps half of them are in the cerebral cortex, with the rest distributed between the cerebellar cortex and the various mini-organs of mid- and hindbrain. Surrounding the neurons are the many glia, all bathed in a solution derived from the cerebrospinal fluid and in contact with a dense mesh of capillaries carrying the richest blood supply of the entire body.

Neurons differ dramatically in form, as Fig. 2.8 in Chapter 2 shows, but their basic functions and chemistry, as identified in a wide variety of vertebrate and indeed invertebrate species, are standard. Not even the most experienced electron-microscopist could distinguish between tiny sections of brain tissue taken from humans, other primates, rodents or birds, except by context. The basic biochemistry and physiology of neuronal function has mainly been derived from work with standard laboratory rodents (rats and mice), while the physiology of axonal transmission is based on now-classical studies, conducted over the 1930s to 1950s, with giant axons derived from the squid.

Neuronal dendrites and cell bodies receive inputs either directly from sensory cells or from other neurons; their axons in turn transmit the summed total of these inputs to other neurons or effector organs. The sites of functional contact between neurons are the synapses. Axons terminate in swellings, synaptic boutons, packed with small vesicles loaded with neurotransmitter. Each neuron may receive inputs from many thousands of synapses along its dendrites, on its dendritic spines and on its cell body, although as many of the synapses may originate from the same neuron, contacts between neurons can be one to one, one to many or many to one (see Fig. 2.5).

At the point at which the synaptic boutons make contact with the dendrite or cell body of the post-synaptic neuron, there is a thickening of the post-synaptic membrane, within which are embedded the receptors for the neurotransmitters. In response to signals arriving down the axon, the vesicles in the synaptic bouton move into and fuse with the membrane, releasing their neurotransmitter, which diffuses through the small gap ('cleft') betwen pre- and post-synaptic sides, binds to the receptor and in doing so changes the balance and flow of ions (sodium, potassium, calcium), at the receptor site, as described in outline in Chapter 2. This in turn either depolarises (excites) or hyperpolarises (inhibits) the post-synaptic membrane. Post-synaptic enzymes then inactivate the neurotransmitter. Excess neurotransmitter can also be removed by being taken back into the pre-synaptic bouton (reuptake) or into the surrounding glial cells. Neurotransmitter release is 'quantal' and depends on the number of vesicles that fuse with the pre-synaptic membrane. If enough quanta are released, the polarising effect at the receptor sites spreads like a wave along the dendrite and cell body to the point at which the axons sprout from it, the axon hillock, where it is summed with all the other inputs from other active synapses on the neuron. The hillock serves as a democratic vote counter of all the synaptic voices reaching it. The electoral judgement is a first-past-the-post method; if the excitatory yes voices are powerful enough to outweigh the no voices coming from inhibitory synapses, they trigger an all-or-none signal which in turn flows down the axon, affecting all that neuron's own synapses. How fast the signal is transmitted down the axon depends, for physico-chemical reasons that do not concern me here, on whether the axon is myelinated (that is, insulated) or not.

In outline this process sounds relatively straightforward. However, there are a large number of factors – both biochemical and architectural – that complicate and enrich it. First, there are many different neurotransmitters, perhaps as many as fifty, some excitatory, some inhibitory. Second, for each neurotransmitter there may be many different receptors, which in turn affect the way that the post-synaptic cell responds to neurotransmitter release. The commonest excitatory transmitter in the brain is the amino acid glutamate.* But there are three distinct types of receptor for glutamate, each of which itself exists in multiple subtypes which have proved happy hunting grounds for neurophysiologists and molecular biologists, who have cloned them and studied their physiological properties in enormous detail. The distribution of receptor types is non-random, varying predictably between brain regions. The effect of glutamate release on the post-synaptic cell depends on which type of post-synaptic receptor it interacts with, although the functional significance of the multiple subtypes, more of which are continually being identified, remains obscure.

To complicate matters further, swimming in the cerebrospinal fluid, the extracellular spaces between the neurons and the cleft between pre- and post-synaptic neurons, are a variety of other secreted substances, known generically as neuromodulators. These are mainly but not only peptides, capable of interacting with receptors on the neuronal surface and at the synapse. They include neurosteroids, hormones such as vasopressin and growth factors such as BDNF (described in Chapter 3) that in the developing brain guide migrating neurons and axons towards their targets. In the adult they mediate neuronal plasticity, remodelling synapses in the light of experience.

The consequence of all this complexity is that whether a signal arriving at the synaptic bouton is effective in triggering a response in the post-synaptic membrane depends not merely on the amount of transmitter released and the sensitivity of the receptors to it but also to the extent to which inactivating enzymes, reuptake mechanisms and miscellaneous neuromodulators are present. As the effects of these modulators may be long-lasting, the present activity of any synapse is also dependent on its past history.

---

* Glutamate also serves as a food flavouring in some cuisines, notably Chinese and Japanese, but, if it is taken in excess in such diets, it can act as a neurotoxin by over-exciting glutamatergic synapses in the brain. A few years ago there was an agitated discussion in the neuroscientific literature about so-called 'Chinese restaurant syndrome' – an unpleasant though brief set of neural sensations resulting from an excess of glutamate flavouring.

Glia

Glial cells are a heterogeneous group, broadly separable into three types with distinct roles: astrocytes, oligodendroglia and microglia. Microglia, the smallest and least common, are scavenger cells, part of the brain's defence system against invasive toxins and viruses. The primary function of the oligos is in myelination, synthesising the fatty sheaths that insulate the longer axons and give the brain's white matter its characteristic appearance. Astrocytes surround neurons and close-pack synapses. They also wrap themselves around the capillaries, so that much of the ionic and molecular traffic between the neurons and the world outside the brain has to pass through them; they thus constitute a protective barrier. However, their cell membranes also contain receptors and reuptake systems for some neurotransmitters, and they secrete a range of neuronal growth factors, so they have a part to play in regulating the synaptic environment, and perhaps other roles too, during development in instructing neurogenesis and in synthesising proteins necessary for neuronal function.[9]

Dynamic architecture

Describing the function of the astrocytes thus makes it clear why it is not merely composition but structure, and relationships that are fundamental to any understanding of the brain. The packing of neurons – their interactions, their interconnectedness – determine who talks to whom. The effectiveness of a synapse, independently of how much transmitter it releases, depends on its address and its geometry. The closer it lies to the cell body of the post-synaptic neuron the more likely is it that its voice will be heard at the axon hillock when the synaptic votes are being counted. If it is too far away, the depolarisation it induces may die away before ever arriving, a wasted vote. Excitatory synapses are more likely to be located on dendrites, inhibitory ones on cell bodies. But – remember Fig. 5.1 – the dendrites themselves are studded with small spines. Some synapses are made on spines, some on the main dendritic shaft, and – again for physicochemical reasons that don't matter to this account – spine synapses have louder voices than shaft synapses. Because neuronal geometry determines connectivity, the morphology of the dendrites themselves, with their many branches,[10] adds to the complexity of the computations they must make of the synaptic voices reaching them, even before their summation at the axon hillock.[11] To add to the complexity, there are not only synapses making contact with dendrites and cell bodies; some actually make connections with other synaptic boutons. Thus an excitatory synapse on a dendritic spine may have sitting on it and controlling its activity an inhibitory synapse from a third neuron. No wonder that this web of interconnections beggars in complexity even the most intricately engraved silicon chip.

Even so, the account I have given is wholly misleading if it implies that the structures and relationships that one can observe in the electron microscope are in any way fixed. The picture of the developing brain in Chapter 3 was one of the steady migration of neurons and glia, their locating themselves in appropriate patterns in the cortex and other brain regions, making and pruning synaptic connections with neighbours and relations. The implication perhaps was that once development was complete, these patterns were essentially stable. Unfortunately, this is also the impression that one might get from the microscope pictures. To make such pictures the brain tissue has indeed to be 'fixed' – dehydrated, chemically stabilised, embedded in plastic, stained to render particular structures visible* and cut into salami slices to view under a beam of electrons. The result is less a snapshot than a fossil, a pitiful remnant of the dynamic living forms from which it was derived.

There are newer, less destructive ways of observing living neurons. They can be grown in tissue culture and photographed under time-lapse. There are even some techniques that enable literal translucent windows to be opened into the brain so that cells can be seen and their activities followed *in situ*.[12] Now neurons can be seen not fossilised, but in real life. The difference is dramatic; even though mature neurons form a relatively stable, non-dividing cell population, their own shapes are not fixed but in constant flux. Under time-lapse the dendrites can be seen to grow and retract, to protrude spines and then to withdraw them again, to make and break synaptic contact. In one study, in the region of the mouse brain that encodes information from the animals'

---

* 'Visualised' as the jargon has it – an interesting concept, for it reminds one just how far the very structures we observe are brought into existence by the techniques we use to observe them. If they weren't fixed and stained then they might not even exist in the form we see them.

whiskers, 50 per cent of dendritic spines persist for only a few days.[13] If this be architecture, it is a living, dynamic architecture in which the present forms and patterns can only be understood as a transient moment between past and future. Truly the present state of any neuronal connection, any synapse, both depends on its history and shapes its future. At this as at all levels of the nested hierarchy of the brain, dynamism is all. The brain, like all features of living systems, is both being and becoming, its apparent stability a stability of process, not of fixed architecture. Today's brain is not yesterday's and will not be tomorrow's.

So perhaps the synapses are the seats if not of the soul then at least of mentation and consciousness? Some years ago the Catholic convert and neurophysiologist Jack Eccles, distinguished for his discoveries of synaptic mechanisms but also a committed dualist, argued that the uncertainties, the indeterminacies, of synaptic action offered an escape from materialism, a literal God of the gaps. He claimed a special brain region, designated 'the liaison brain' in the left hemisphere, as the point at which the soul and hence the deity could intervene and tinker with neural mechanisms.[14] Eccles's views, at least on this, are scarcely taken seriously these days, being replaced by Roger Penrose's microtubular mysticism. Today the dominant trend amongst leading neuroscientists is decidedly reductionist in its insistence on molecular explanations.[15] These are powerful voices, but I will continue to insist on the distinction between enabling and either determining or even 'being the same as'. Just as mentation and consciousness are not reducible to biochemistry nor are they collapsible to individual synapses or individual neurons.

Enter the pharmacologists

Focusing on synapses and their interconnections suggests – and there are many neuroscientists who would go along with this – that they are the central actors in the drama of the brain and hence of mentation. The title of the recent book by neurobiologist Joe Ledoux – *Synaptic Self*[16] – neatly encapsulates this synaptocentric perspective, and indeed it has a lot going for it. For decades now the burgeoning pharmaceutical industry has focused on the synapse and its neurotransmitters as the site for intervention into brain function. The overwhelming majority

of drugs formulated to alter mental states, as treatments for both neurological disorders like Alzheimer's or Parkinson's disease, and psychiatric
diagnoses such as depression, anxiety or schizophrenia, are designed to
interact with neurotransmitter function. (I'll have more to say about this
in later chapters.) Molecular mimics of neurotransmitters can supplement supposed deficiencies, as in the use of L-dopa to compensate for
the loss of the neurotransmitter dopamine in Parkinson's disease. They
can interfere with the enzymes that break down the neurotransmitter,
as with drugs like rivastigmine or aricept that diminish the destruction
of the neurotransmitter acetylcholine by the enzyme acetylcholinesterase
in Alzheimer's disease. They can block the removal of the neurotransmitter by reuptake mechanisms, as with the SSRIs – specific serotonin
reuptake inhibitors – of which Prozac is perhaps the best known. They
can be entirely artificial molecules which compete with naturally occurring neurotransmitters in binding to one or more of their receptors, like
Valium (benzodiazepine), which interacts with receptor sites for the inhibitory neurotransmitter GABA. Also, of course, many so-called recre-
ational drugs, from nicotine to LSD, act chemically at specific synapses.

Building ensembles

The effects of such psychoactive chemicals are of course not confined
to humans – they tend to affect laboratory animals in ways that appear,
at least externally, analogous to their effects in humans. Indeed, this is
how many have been discovered and tested before being brought into
clinical trial. This once again emphasises the continuities between
human and non-human brains at the biochemical and cellular level. It
is as if, evolution having invented the neuron and its synapse as the
unit structure for brain work, there was little point in altering it subsequently. What begins to matter as one moves from species to species,
brain to brain, is the ways in which these units are organised. Neither
neurons nor synapses are isolated monads. They are components within
communicative structures. Just as their internal biochemistry must be
seen in geometric context, so too it is not merely the individual neuronal
architecture but its relation – topological and dynamic – with others.
Who its neighbours are, who speaks to it via its synapses, and to whom
in turn it speaks, determine the role of any neuron in the functioning

of the organism. The brain, as I keep emphasising, is a collection of mini-organs. Those buried deep in mid- and hindbrain and sometimes, if confusingly, called nuclei, consist of dense masses of neurons, numbering tens or even hundreds of millions, connected with each other by way of interneurons, and reaching out via longer, myelinated routes to other ensembles elsewhere in the brain. And then of course, there are the vast concentrations of neurons in the cerebral and cerebellar cortices, each showing a degree of functionally and anatomically modular organisation. Neurons within each module speak to one another, there are feedback and feedforward interconnections between the modules, and there are projection neurons, with long myelinated axons which reciprocally connect other brain nuclei with specific cortical regions. For well over a century, generations of neuroanatomists and neurophysiologists have been painstakingly tracing the wiring diagrams that link all these regions. The pathways can be identified structurally by micro-injecting a dye into a neuronal cell body and tracing the flow of the dye down the axon to its synaptic terminals – and sometimes even across into the post-synaptic neuron. Or they can be tracked functionally, by stimulating one neuron and recording from another distal to it. The time taken for the second neuron to respond can be used to calculate the synaptic distance between the two – that is, how many synapses must be crossed to arrive.

The wiring diagrams that are derived from such studies are an engineer's joy and nightmare in their complexity. Whatever might be the case amongst the global human population, there are certainly less than six degrees of separation between any two neurons in the human brain. Certainly the complexity is sufficient for any reasonable reductionist sensibly to seek to locate the brain causes, or at least correlates, of mentation and action within such complex ensembles. Maybe we are at the right level at last?

One of the simplest examples of how a wiring diagram can illuminate functional mechanisms comes from the mapping of the cerebellum, the second largest mini-organ in the brain after the cerebral hemispheres. Its structure is so regular and apparently so simply related to function as to lead those who pioneered its study to co-author a book entitled *The Cerebellum as a Neuronal Machine.*[17] Functionally, the cerebellum is primarily concerned with the fine tuning of motor output initiated in the motor regions of the cerebral cortex, preserving balance and regulating eye movements. Apart from the glia, the cerebellum contains five classes of neurons, four inhibitory (stellate, basket, Purkinje and Golgi) and one excitatory (granule cells). There are two inputs, mossy and climbing fibres, and one output, from the Purkinje cells. Fig. 6.1 shows the architecture. The most striking feature is the array of large Purkinje neurons, arranged in a layer parallel to the cerebellar surface, each with a huge dendritic tree running in the vertical plane towards the surface. Below the Purkinje cell layer there is a layer of granule cells, whose axons run parallel to the cerebellar surface and at right angles to the plane of the Purkinje cells, so that each granule cell, excited by input from the mossy fibres, in turn makes excitatory synaptic contact with many Purkinje cells. The climbing fibres by contrast make powerful excitatory synaptic contact directly on the Purkinje cell bodies, each climbing fibre in contact with up to ten Purkinje cells. The stellate,

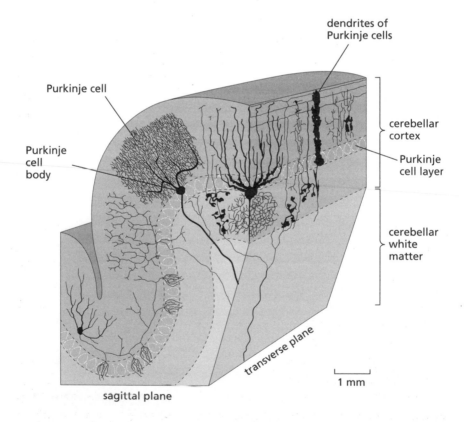

Fig. 6.1 *Cerebellar architecture.*

basket and Golgi neurons all make inhibitory synapses on the Purkinje cells. The single output from the system, the Purkinje cell axon, is in turn inhibitory of lower brain structures more directly concerned with motor outputs.

This peculiarly precise geometry indeed makes the cerebellum appear almost machine-like, serving something of the same function as a governor on a steam engine, damping down excessive movement. It might seem that here at least was a rigidly specified structure, which showed little scope for plasticity. Yet there is more to the cerebellum than this description would imply. It has a role in planning movement and evaluating sensory information for action, and hence it, like many other brain structures, is able to learn – that is to modify output adaptively in response to experience. The cerebellum is engaged in modifying certain simple reflexes, such as the way in which we and other animals blink our eyes in response to a puff of air. Humans, and other animals, can be taught to modify this reflex – for instance, after a few trials in which a buzzer or flash of light is presented a few seconds in advance of the air puff, we blink to the signal rather than await the puff. The circuitry responsible for this reflex also has a connection to the cerebellum, and it is the synapses there that respond to the signal and 'learn' to trigger the blink in advance of the air puff. However, just where in the cerebellar architecture the learning synapses lie remains a matter of some experimental disagreement.[18]

If the cerebellum seems simple but masks complexity, the problem of understanding the visual system is far greater. As I described in Chapter 2, signals from the retinae flow via the optic nerves to the lateral geniculate nucleus in the thalamus and from there to the visual cortex. Considerable sorting and data processing has already been achieved before the cortical level. There are patients who have suffered brain damage in which the visual cortex has been destroyed and they are therefore functionally blind, unaware of visual stimuli even though these are still being processed at lower levels. Such patients show a phenomenon that has been called blindsight; even though they deny that they can see, if they are asked to make a guess as to, for instance, the motion of an object in front of them, they are frequently correct.[19]

Within the primary visual cortex, with its characteristic six layers from surface to the white matter below it, specific cells respond differently to different stimuli. The classical recording studies were made by Torsten

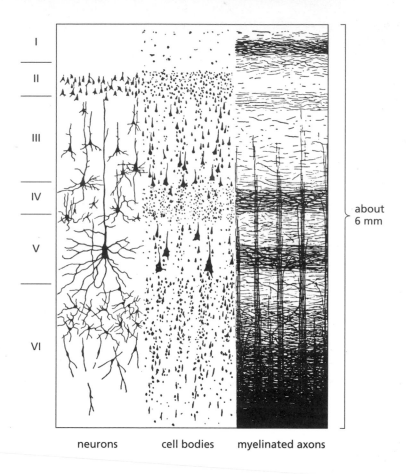

I

II

III

IV

V

VI

about
6 mm

neurons   cell bodies   myelinated axons

Fig. 6.2 *The six layers of the visual cortex, seen using three different types of stain to show different neuronal features.*

Wiesel and David Hubel from the 1960s on. Their findings are illustrated in Fig. 6.3. Think of a cube of cortex. Within the cube, some cells (simple cells) respond best to lines of particular width, others (complex cells) to the ends of lines or corners. Some respond to vertical, some to intermediate and some to horizontal lines. Some respond to inputs from the left and some from the right eye. The cube is organised so that the simple cells tend to be in the lower layers, the more complex ones higher. Inputs from left and right eye alternate in one horizontal plane of the cube, cells with different orientation specificities in the other. In other regions of the cortex, the colour-detecting cells are arranged in short

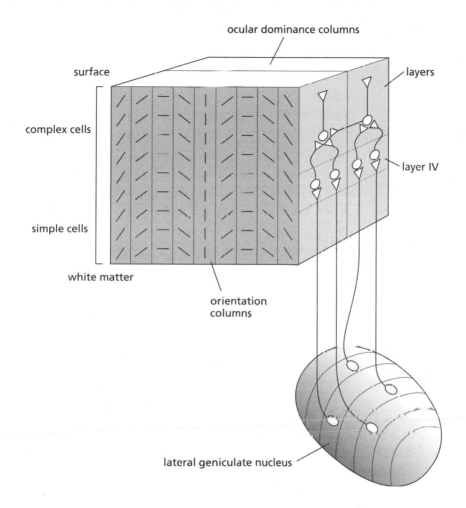

Fig. 6.3 *Three-dimensional block of visual cortex showing input from the lateral geniculate and arrangement of simple and complex cells, orientation columns and ocular dominance columns.*

round columns, or, when seen from above, as 'blobs'. More recent analysis by Semir Zeki has culminated in a vision of the visual cortex as parcelled out into at least thirty different modules, each independently concerned with analysing one aspect of visual information, such as colour, form, motion, direction and angularity.[20] The obvious question following the discovery of an organisation of this type is: how are all these different modes of analysis of visual information integrated to provide what normal people usually experience – a unified picture in which colour, shape and motion are all simultaneously identified and associated with specific objects in the visual field? This has become the crucial theoretical problem for neuroscience.

Cartesian homunculi and the binding problem

The problem extends far beyond just the unifying of visual perception, though. The cortex is parcelled out into specific regions associated with registering and interpreting sensory inputs, initiating motor activity, engaging speech and so forth, and all have to be co-ordinated. Each region shows some functional topology, but how are these individual regional functions to be integrated? The old view was that all the individual mini-regions and modules reported upwards, so to speak, to some central co-ordinating area located in the cortex somewhere in the frontal lobes. And there at the top of the command chain would sit what Daniel Dennett has memorably called the Cartesian homunculus,[21] assessing information and issuing orders. But closer consideration of such an idea makes clear that it simply won't do. There is nowhere in the brain a site at which neurophysiology mysteriously becomes psychology. The logical objection is that to posit a homunculus (your brain's office manager, as my childhood encyclopaedia described it) transfers the problem of how the whole brain works to that of how a theoretical mini-brain with its own mini-office manager inside works. The prospect of an infinite regress of smaller and smaller homunculi simply won't do. Also, more importantly perhaps, if there were such an area, neuroscience has conspicuously failed to discover it. As Gertrude Stein put it in a rather different context, there is simply no there there. There is no super-boss in command. Nor can there be. It took the experience of Stalinism in the former Soviet Union and Fordism in capitalism to show that command economies don't work. But nor can neural systems in which information passes upwards and inflexible commands return downwards. Instead, the brain operates like a classical anarchistic commune in which the semi-autonomous work of each distinct region contributes harmoniously to the whole: from each region according to its abilities, to each according to its needs.

The idea of some central command system or, to adopt computer speak, a central processor, dies hard though. Fig 6.3 shows a classical

Fig. 6.3 *The Cartesian model of the automatism that determines the withdrawal of a foot from something that burns it. From* L'Homme de René Descartes, *Paris, 1664.*

diagram: Descartes' vision of a person's response to a flame approaching his foot. In the Cartesian model, messages were supposed to pass up the spinal cord to some hypothetical pain centre in the brain, and from there down, instructing the foot to move. But, as the neurophysiologist Pat Wall pointed out,[22] although the 'pain fibres' in the spinal cord are indeed there and can be studied in detail, the hypothetical pain centre simply doesn't exist; rather, the experience of pain is a device for mobilising action in a variety of brain regions so as to diminish danger. This is why, as Wall goes on to discuss, terribly injured soldiers in the battlefield may not experience pain at all, as it is counterproductive to their survival – until they are rescued and on the way to the relative safety of hospital where the pain may be permitted.

If the presumed 'higher centre' which provides the unitary sense of being a person in conscious control of one's senses and actions simply doesn't exist, how does information become transformed into perception, become meaningful? Think of seeing a red car moving along the road in front of you. One visual module says red, another detects a car shape, a third motion, a fourth directionality, perhaps a fifth speed. If there is no unifying and interpretative homunculus to receive and pull all the separate bits together, how is it to be done; how can experience be unified? The answer has to be that it happens in the visual cortex itself; multiple interconnections link the separate modules, integrating their separate analyses; the flow of signals between them is the mechanism of perception. This, however, also means that the visual cortex cannot just be passively receiving the information; it must be actively comparing it with past experience – to recognise the car as a car for instance. That is, some forms of memory must be either located within or accessible to the visual cortex.

That the visual cortex is more than a passive analyser is shown quite neatly in some recent experiments of my own using MEG. We[23] were using supermarket shopping choices as a way of studying real life memory. Subjects were shown a series of pictures each offering a choice of one of three potential purchases – for instance three brands of coffee or beer – and asked to choose the one they most preferred by pressing a key. In a control experiment, they were shown the same images again and asked, also by keypress, to indicate which was the smallest/shortest. It takes people about two and a half seconds to make their choices and press the key. Over that time the MEG shows a wave of activity crossing a variety of cortical regions. The first significant signal appears, scarcely surprisingly, in the visual cortex, about 80 milliseconds after the images appear on the screen. What is interesting in the context of the present discussion is that the visual cortex signals are much stronger in the preference choice than when judging which is the shortest image. That is, the same images evoke different visual cortex responses depending on the context in which they have to be assessed. The cortex is not merely receiving inputs but actively involved in interpreting context, turning input into perception.

Both the example of the red car and the MEG experiment open up the wider ramifications of the problem of how disparate brain processes can give rise to integrated, unified experience. It is not sufficient for the visual cortex to link the visual aspects of the experience; cars make a noise, so auditory input is involved; if the car is close and you are crossing the road, you might experience alarm and fright, and need to move suddenly to avoid being hit. Signals involving alarm and fear arrive in the brain via the amygdala; motor movements involve the topographically organised motor cortex, basal ganglia and cerebellum; all must be co-ordinated within the unitary experience involved in seeing, identifying, the red car as a potential source of danger, of planning appropriate action to avoid it, and, finally, acting – perhaps jumping out of the way.

In the MEG experiment, we observed a wave of activity traversing the cortex in the first second after the appearance of the images; following the visual cortex, the left inferotemporal region, known to be associated with semantic memory lights up; a little later, especially if the choice is a hard one, Broca's area, as the product names are silently verbalised; finally, if and only if a preference choice is made, a region in the right parietal cortex, which, we believe, is concerned with integrating emotionally relevant conscious decisions – at least, damage to this region can affect such decision-making, according to Damasio.[24] Thus the apparently seamless process of observing, evaluating and choosing between images based on past experience involves multiple brain regions, not necessarily connected via simple synaptic pathways. Achieving such coherence is called 'binding', and, as I have said, is the central theoretical problem for twenty-first-century neuroscience.

Whilst the cortex is indeed parcelled out into numbers of functionally distinct regions during development, within these regions neurons concerned with particular aspects of analysis or instruction are topographically organised; as in the visual cortex, these maps are themselves highly dynamic, remodelling themselves in response to experience.[25] Even though there are critical developmental periods in which such responses are most dramatic – as in the examples of the visual cortex, or language learning in the previous chapter – even adult maps are unstable; use can expand cortical maps, disuse contract them. Learning complex digit movements involving say, left thumb and forefinger, expands both their sensory and motor representation. Although loss of a limb leaves its representation in the brain intact for a period, leading to the experience of possessing a 'phantom limb',[26] disuse can leave a space in the map which can be taken over to expand the representation of other more actively engaged motor or sensory systems.

Multiple forms of signalling

A single axon can convey a single message: its all-or-none response to the summation of signals arriving at the axon hillock. However, an axon firing at a rate of once per second conveys very different information from one firing at a rate of forty per second (40 hertz). Information is embedded in the frequency with which an axon fires, but also in the rate of change of that frequency – a sudden speed up or slow down itself is informative. But axons are never alone; modules contain tens or even hundreds of millions of neurons acting in concert – otherwise the MEG would never be able to pick up their activity. Nerve fibres running between the different brain regions are bundles of axons, which respond to the activity within the module; they too may be firing in synchrony or out of phase with one another. Synchronous firing conveys different information to the receiving neurons than out-of-phase firing.

Since the 1920s it has been known that electrodes placed on the scalp record complex wave patterns, the electro-encephalogram. As the implications of the EEG became apparent, many physiologists were convinced that, if the patterns could be interpreted, they would provide a direct correlate of mental processes, revealing hitherto unsuspected aspects of brain activity.[27] The wave forms alter between sleep and wakefulness, and change when a person is asked to undertake focused activity. The early hopes proved unfounded; more reductionist neurophysiological approaches implied that they were nothing but epiphenomena, causally irrelevant. Today spontaneous EEGs and, more importantly, what are called evoked responses – that is, the EEG signals arising when a person is asked to carry out some task – are once more back in fashion, along with their correlated measure, the MEG. MEG signals and evoked potentials are measures of the correlated activity of hundreds of thousands if not millions of neurons. They reflect mass synaptic activity. Furthermore, this activity is often synchronous and oscillatory. Focused neural activity such as task solving generates a particular rhythm, a pattern of 40-hertz oscillations which can be detected in brain regions which may be synaptically distant but engaged in complementary aspects of the same task – linking auditory, visual and memory-related stimuli for instance. The neurophysiologist Wolf Singer, amongst others, has explored the implications of this 40-hertz oscillation and has suggested that it is this that serves to bind together different brain regions.[28] That is, one must consider the workings of the brain not merely in terms of three-dimensional spatial anatomical connectivity but in the time dimension as well. Cells that fire together bind together. Once again, the message is that brain processes depend on history as well as geography.

A further complexity. Even when the temporal as well as spatial contiguity is taken into account, each axon connects with specific target cells. It is a labelled line. But there is an additional phenomenon that comes into play when considering the combined activity of masses of cells, as the neurophysiologist Walter Freeman has pointed out. To form a complex system, neurons must be semi-autonomous, each with weak interactions with many others, and show non-linear input-output relations. Such systems are open, forming patterns through their collective activity that transcend the cellular level, ceasing to be local. The neurons cease to act individually and begin to participate as a group in which it is the population rather than the individual cell that becomes important, generating stable states that are best understood in terms of chaos theory, with current sources and sinks flowing through the neuronal population.[29] Elucidating the maths of such chaotic dynamics would take me outside my brief here and indeed, to be honest, outside my competence; the point that concerns me is that this population-based property of neuronal masses represents yet another level within the nested hierarchy which generates the brain processes associated with thought and action.

Memory and the paradox of levels

So, molecules, macromolecules, neuronal and synaptic architecture, ensembles, binding in space and time, population dynamics; all clearly contribute to the functioning of the perceiving, thinking brain. At what level does information processing become meaning, awareness, even consciousness? I asked at the beginning of this chapter, and answered: at all levels and at none. The phenomena of learning and memory beautifully illustrate the paradox.

Memory, my own central research topic, is our most defining property; above all it constitutes our individuality, provides our life's trajectory with autobiographical continuity, so that we seem to be able, even in the eighth and ninth decades of our life, to recall episodes from our childhood. How can this stability be achieved, even when every molecule of which the brain is composed has been synthesised, broken down and replaced by others more or less identical many trillions of times? Somehow, it is assumed, experience, learning, must be stored in the brain, to be accessed either voluntarily or involuntarily at some later date. Apart from a brief flirtation in the 1960s with macromolecules such as protein or nucleic acid as memory stores,[30] the dominant assumption amongst neuroscientists follows a hypothesis made originally by the psychologist Donald Hebb in 1949:[31] that novel experience (whether incidental or specifically learned) resulted in changes in synaptic connectivity, strengthening some synapses and weakening others, so as to create new pathways amongst some sets of interneurons, which represents the memory in some way, perhaps analogous to the trace on a CD or magnetic tape.*

There is overwhelming evidence from animal experiments that novel experience, such as training rats, mice or chicks on simple tasks, results in such synaptic changes. In my own experiments with young chicks, a novel learning experience results in increased neural activity in a specific region of the chick brain, release of neurotransmitter, activation of particular receptors, and, over a period of some hours, synthesis of new membrane proteins and measurable changes in synapse size, structure and even number. Similar sequences have been found in other laboratories and species, and even modelled physiologically with slices of brain tissue. So was Hebb right: is memory biochemical and synaptic?

But this is where the paradoxes begin, for neither in the chick nor in mammals does the memory 'stay' where the initial synaptic changes occur. If the specific region of the changes in the chick brain is removed a few hours after the learning experience, the memory, surprisingly, is not lost. In mammals, the key 'entry points' for learning and memory formation are the hippocampus and (for emotional memories) the amygdala. But, as with the chick, the memories do not remain there locked

---

* Later theorists have modified this idea somewhat, suggesting that rather than pathways there are new patterns of stronger synapses within a network of neurons. Modellers interested in artificial intelligence call such patterns, which can be computer-simulated, 'neural nets'. Whether such nets exist in real brains remains debatable. I have discussed this in *The Making of Memory*, but whether we are dealing with pathways or patterns is not germane to the main issues I am concerned with here.

into some fixed store, but become much more diffusely distributed in the brain. The classical case is of a man, known only by his initials, HM, who was operated on in the 1950s to treat his epilepsy. Regions of his hippocampus and surrounding tissue were removed. The epilepsy diminished, but HM was left with a terrible memory deficit. He could remember well events in his life up to the time of the operation, but nothing subsequently. New experiences, learning, could no longer be retained for more than a few moments. Fourteen years after the operation, according to Brenda Milner, the psychologist who has given him the closest study over the years:

> He still fails to recognize people who are close neighbors or family friends but who got to know him only after the operation . . . Although he gives his date of birth unhesitatingly and accurately, he always underestimates his own age and can only make wild guesses as to the [present] date . . . On another occasion he remarked 'Every day is alone by itself, whatever enjoyment I've had, and whatever sorrow I've had.' Our own impression is that many events fade for him long before the day is over. He often volunteers stereotyped descriptions of his own state, by saying that it is 'like waking from a dream'. His experience seems to be that of a person who is just becoming aware of his surroundings without fully comprehending the situation . . . H.M. was given protected employment . . . participating in rather monotonous work . . . A typical task is the mounting of cigarette-lighters on cardboard frames for display. It is characteristic that he cannot give us any description of his place of work, the nature of his job, or the route along which he is driven each day . . .[32]

So, the hippocampus is necessary for learning new information, but not for 'storing' it subsequently. Imaging studies show other regions of the brain, such as the left inferotemporal cortex, as being active during memory retrieval. But so, as in the experiment I described above, is the visual cortex when a person is instructed to make a visual choice based on memory. There are other paradoxes; recalling something is an active, not a passive process; it isn't simply retracing a pathway made by Hebb synapses. Indeed there is good evidence that the act of recall, retrieval, evokes a further biochemical cascade, analogous to, though not identical with, that occurring during initial learning.[33] The act of recall remakes a memory, so that the next time one remembers it one is not remembering the initial event but the remade memory from the last time it was invoked. Hence memories become transformed over time – a phenomenon well known to those who study eyewitness accounts of dramatic events or witness testimony given in court. Once again, memories are system properties, dynamic, dependent, for each of us, on our own unique individual history. What they absolutely are not is 'stored' in the brain in the way that a computer stores a file. Biological memories are living meaning, not dead information.

But even this is over-simple. To talk about 'memory' is a reification, a matter of turning a process into a thing. It implies that there is simply one phenomenon called memory. Psychologists recognise a multidimensional taxonomy of memory. There is a distinction between knowing how and knowing what. Remembering how to ride a bicycle is different from remembering a bicycle is called a bicycle. Remembering the names and order of the days of the week is different from remembering what I did last Wednesday; and remembering something transiently such as a string of eight numbers is different from recalling an often used telephone number. Procedural and declarative, semantic and episodic, short- and long-term, working and reference memory, remembering faces and remembering events . . . The list extends indefinitely. Are there 'memory systems' in the brain, as some would argue, citing different memory stores for object names, text and auditory words?[34] If so, perhaps the different systems even compete with one another. Or is the idea of multiple memory systems a figment, conjured into existence by the nature of the tasks subjects are given?[35] What is clear is that however much we may have discovered about the neural processes involved in learning, it is still quite unclear how the recall process works. Clearly, remembering is an active, not a passive event, and draws on a variety of cognitive and affective processes.

Imaging studies suggest that the frontal lobes are involved in the strategic task of hunting for and assembling the memory, drawing on semantic 'stores' in the inferotemporal cortex. Perhaps the parietal may be involved in 'ageing' the memory – that is locating time in a person's past experience,[36] a problem that has vexed philosophers from Aristotle to St Augustine. The psychologist Endel Tulving has gone so far as to argue that 'memory' as such does not exist in the sense of being instantiated in brain structures; rather it is actively called into being in the process of recall – or, as he terms it, 'ecphorised'.[37]

Functional systems

Some of these modern views were presaged many years ago, in the 1940s and 1950s, by the Soviet Russian neuroscientist Peter Anokhin, once a pupil of Pavlov, who developed what he called a 'functional system' theory.[38] He and his colleagues recorded from a variety of cells in different regions of the rabbit brain, and noted that particular patterns and combinations of firing of cells in very different brain regions occurred specifically when the animal was in a particular place and engaged in a particular activity – say the right hand corner of the cage and eating a piece of carrot. What linked the neurons together was their joint engagement in a particular goal-directed activity, rather than being stable passive sites simply 'representing' the outside world.[39] Soviet neuropsychological theories have not on the whole been well received by Western science, and Anokhin's has been no exception. They have, however, been slowly rediscovered and reformulated, as when John O'Keefe and Lynn Nadel discovered that the hippocampus contains what they called place cells – cells that fire at particular map references in a rat's environment and in conjunction with specific activities. In 1978 they conceptualised this in a ground-breaking book, *The Hippocampus as a Cognitive Map*.[40] But the crucial point, which brings together in their different ways Singer and Freeman, Anokhin, and O'Keefe and Nadel, is that systems do not exist in the brain in abstract;[41] they are called into play by actions, and are as transient and dynamic as the actions themselves.

Brains in bodies

Confession time. Many years ago in the green arrogance of my youth I wrote a book called *The Conscious Brain*.[42] Sharper but kindly critics acclaimed my title as a deliberate paradox. But 'the conscious brain' isn't an oxymoron, it is merely a starting point. As I have emphasised again and again, brains are not independent of bodies. Of course, they are dependent in their need for a regular supply of glucose and oxygen, and they are dependent on the sense organs located through the body for providing their inputs on which to act, to 'instruct' muscles, endocrine organs, immune systems, to express 'will'.

But body signals to brain in other ways too, notably through hormone systems and especially in registering affect – concern, stress, fear, alarm, but also contentment and joy, euphoria. These are emotions, which in the context of human consciousness become feelings, as Damasio puts it. Two regions of the brain serve particularly as entry points for such hormonal inter-actions. Fearful experiences trigger the production of the 'flight and fight' hormone adrenaline, from the adrenal glands. Adrenaline is taken up into the brain where it interacts with receptors in the amygdala, a key part of the brain's emotional circuitry. In an ingenious set of experiments, Larry Cahill and James McGaugh showed, first, that people remembered emotion-laden information such as fearful stories better than more purely cognitive ones. They then told the same stories to subjects given propanolol, a drug that blocks adrenergic inputs into the amygdala. Under these conditions the fearful stories were remembered no better than the cognitive ones.[43] There are two lessons from this experiment. First, emotional memory is more powerful than purely cognitive; and second, that body and hormonal state affect how well something is remembered.

The steroid hormones, oestrogen and testosterone, and corticosterone, tell a similar story. I've already mentioned that the hippocampus is rich in receptors for corticosterone, the 'stress' hormone.* There's a fine balance in its effects on the brain between too much and too little corticosterone. The neuroendocrinologist Bruce McEwen has shown that a chronic overproduction of corticosterone over the life cycle accelerates the death of hippocampal neurons, a phenomenon of life-time stress that he calls allosteric load.[44] On the other hand the optimum level of corticosterone can enhance learning and memory retention. In our own experiments with chicks, the birds are normally kept in pairs as they dislike being solitary. If they are trained as a pair and then imme-diately separated, however, their blood steroid level increases and they

---

* There are actually two types of hippocampal steroid receptors, and the steroids themselves have two biochemically different effects; one is an interaction with receptors on the cell surface, the other involves entry into the cell so as to affect gene expression directly.

show better memory retention.* Blocking the steroid receptors in the chick brain results in amnesia for a recently learned task.[45]

Brains and bodies in the world

But of course, brains are not merely in bodies. The individual organism, the person, is in the world. Brains and bodies are open, not closed systems, in continuous interaction with the external material, biological and social worlds. It is this which explains why, with apologies to Emily Dickinson, I want to argue that the mind is wider than the brain. Over the last decade or so there has been a flood of books, conferences and journals intended to bring neuroscientists and philosophers together to discuss the nature of consciousness. Can consciousness be reduced to a brain process, or at least explained in terms of neural mechanisms? The opening paragraphs of this chapter have summarised the types of claim being made. Consciousness 'is', according to some neuroscientists, the fleeting binding together of ensembles of neurons, a moment in time, ever receding into the past, ever moving forward into the future. One particularly egregious simile even offered a comparison between consciousness and the workings of an electrical dimmer switch, brightening as more neurons were engaged, darkening when fewer were active. Consciousness is an evolved brain property, dependent on particular brain structures, suggests Antonio Damasio, in his heroic attempts to deduce the neuroanatomy of feelings, and self-awareness.[46] 'Everybody knows what consciousness is; it is what abandons you every night when you fall into dreamless sleep and returns the next morning when you wake up,' claims psychiatrist Giulio Tononi.[47] Such a narrow definition suits philosophers, who can then distinguish between subjectivity, first person understanding, only accessible to the conscious individual, and objectivity, third person observation of that individual's actions and expressions. Philosophers and neuroscientists can then agree to divide the field between them.

But wait. The writer David Lodge has shown how much richer a novelist's understanding of such 'inner processes' is than this.[48] Think

---

* Interestingly, the steroid level is higher in separated young male chicks, which are more fearful, and as a result, their memory retention is better than that of the females.

for example of the famous Molly Bloom soliloquy in James Joyce's *Ulysses*. And as the sociologist Hilary Rose has pointed out,[49] there are many other understandings and uses of the term consciousness; there is Freudian consciousness, with its murky unconscious world of desires and fears. There is social consciousness, class, ethnic or feminist consciousness, the recognition of having a standpoint from which one can interpret and act upon the world. All these understandings are lost in the impoverished worlds shared by these philosophers and neuro-scientists, who reduce such varied worlds to that of being aware, being awake, being unanaesthetised. Being conscious is more than this; it is being aware of one's past history and place in the world, one's future intents and goals, one's sense of agency, and of the culture and social formations within which one lives. Today's consciousness is not the same as that of a Victorian gentleman like Charles Darwin, or of a Greek philosopher like Plato. Nor was Darwin's consciousness the same as that of a Victorian mill hand, or Plato's equivalent to a Greek slave. To be sure, brain and body enable consciousness, which is one reason why Sue Savage-Rumbaugh's consciousness is not the same as Kanzi's, or mine the same as my four-year-old grandchild. For that matter, at age sixty-five in 2004 my own consciousness is not the same as that I claimed as the writer of *The Conscious Brain* more than thirty years ago.

Of course, I am not the only neuroscientist to argue this. Indeed, a whole new field is emerging: the cognitive neuroscience of human social behaviour – but it will only be made possible in collaboration with psychologists, anthropologists, ethologists, sociologists and philosophers.[50] The brain is indeed specifically adapted to this social openness; for instance, there is a class of neurons ('mirror neurons') that specifically fire when an individual imitates the actions of others, and those that are tuned to register other people's emotions and intentions[51] or infer intention from action.[52] There are brain systems (empathic systems) that respond to observing another's pain.[53] There are widely distributed neural systems involving both cortical and deep brain regions involved in having what some have called a 'theory of mind'.[54] Such a theory – a recognition that there are other than one's own mind in the universe – is argued to be essential for such openness. Thus, to say that brain enables consciousness is not to make a crude split between the neurological and the psychological, the biological and the social, or to separate, if that were possible, the phenomenon of consciousness from the content of consciousness.

There can be no consciousness without content; indeed it is constituted by its content, and its content is not merely of the moment but of all past moments in the history of the individual. It is thus an emergent property, not to be dichotomised, indissolubly historically located. It exists in sets of relationships, between the person and the surrounding world, irreducible to mere neural mechanism but not a mysterious ghost in the machine either. It is subject to scientific investigation, but not to be encompassed by the methods of neuroscience with our imaging devices, electrodes and medicine chest of psychoactive drugs.

This brain then, is that marvellous product and process, the result of aeons of evolution and for each human adult decades of development, the necessary organ of consciousness, thought, memory and identity, which modern neuroscience is beginning both to describe and explain. What the potential consequences of such descriptions and explanations, such electrical and chemical probes into our innermost past are beginning to make possible, I will try to assess in the rest of this book. But not before I bring the evolutionary and developmental lifeline of the brain to its quietus, and thus complete this particular life-cycle.

CHAPTER 7

Ageing Brains: Wiser Minds?

Lifespan and the paradoxes of ageing

'THEY SHALL GROW NOT OLD, AS WE THAT ARE LEFT GROW OLD' intone priests and padres at funeral services. Poets have said it, none more powerfully perhaps than Shakespeare, whose sonnets muse over the decline of youth and the inevitability of ageing, or Marvell, whose lament at finding that 'At my back I still do hear / Time's winged chariot hurrying near' echoes down the ages. There is a time for living and a time for dying; the life cycle has an inevitability which means that one of the few things certain in life is that we will all die. Of course there are those that offer hope. Victorian gravestones are full of euphemisms – 'passed away', 'fell asleep', 'in the sure and certain knowledge of the resurrection'. Today these pious sentiments are being replaced by hoped-for technological fixes, pills that might offer the modern version of the alchemists' elixirs of life, of eternal youth, the escape from Time's chariot, a prospect mordantly anticipated back in the 1930s by the visionary Aldous Huxley in *After Many a Summer*. In that book it is extracts of raw carp intestines that do the job, but the resulting longevity is coupled with a reversion to a pre-human ape-like existence where only copulation staves off the drearily eked-out life beyond the allotted biblical time span.

Today boldly prophet/profit/eering entrepreneurs offer their customers, for a fee, the prospect of cheating death entirely; of having their bodies – or at least their heads – deep frozen until such time in the indefinite future when science may have developed techniques of thawing and rejuvenating. As if a frozen brain, in the stasis of death, could by thawing recover the dynamic of the history that made its owner what he or she once was in life . . .

In an only slightly less huckstering mode, a US memory researcher with a new product on the market once gave a press conference claiming to be able to provide 'a seventy-year-old the memory of a twenty-year-old' and saw his company's share prices rise dramatically. In 2002 *Forbes* magazine, that harbinger of good news about the wealthy and about-to-become wealthy, offered its readers profiles of two other memory researchers working on pills that they proposed would function as 'Viagra for the brain'.[1] Such claims may well also serve as erectile for stock-market values, but they leave open the questions both of technical feasibility and desirability. These and other future technologies will form the focus of the rest of this book; my goal in this brief chapter is to bring the life-trajectory of brain, mind and personhood which has been my theme so far to its present conclusion.

Prolonging the biblical time span comes up against a biological reality that it is hard to circumvent. It is important to distinguish life span (technically, the age of death of the longest-lived member of a species) from ageing (whether there are age-related decrements or changes in biochemistry, physiology and functioning). Death, at least, is relatively unambiguous, and for most multicellular organisms death is no more and no less part of the developmental life cycle than is birth. If you really don't want to die, your best chance would be to abandon sex and take up reproduction by splitting or budding, like bacteria or amoeba. The daughter cells produced in such a way are rejuvenated, and in principle immortal unless destroyed by predators, poison or inclement environments. However, this option for immortality eludes most sexually reproducing beings like ourselves. There are exceptions. There are some animals – particularly those that grow indefinitely large and have no necessary maturational size, like carp (which is why Huxley opted for carp intestines as his elixir), sharks, or the Galapagos tortoise – that have no fixed life span and do not discernibly age, living until culled by predators or accident.

There is good evidence that life span is species-specific, although one of the several residual mysteries is the considerable difference in longevity between species with seemingly rather similar life styles. In general, however, longevity seems to relate in part to size – for example, whereas mice may live for a maximum of about three years, elephants can live for seventy. The 'natural' maximum human life span is said to be 112 (although the recent death of a French woman aged over 120 may extend this)[2] – some way above the biblical limit of four score (discounting Methuselah and his fellow ancients). There are arguments from both physiology and evolutionary theory why this should be the case, at least amongst mammals. The smaller the mammal, the relatively faster its metabolic rate needs to be; rapid metabolic rate, rapid maturation and capacity to reproduce go together; larger size and slower maturation require a longer life span. However, there is no simple relationship. Parrots are notoriously long-lived. In some short-lived organisms such as *Drosophila*, mutations have been generated that increase longevity, whilst some *C. elegans* mutants have more than double the ten-day life span of the wild-type worms.

Whether one ages or not over one's life span is a separate matter. Ageing, like all other aspects of living, is not a biological, but a biosocial phenomenon, as all sensitive gerontologists will recognise, and about which many have written.[3] In older age, one can no longer do the things one achieved with ease as a younger person, whether it is vaulting over a low fence or reading small print without glasses. The older one is, the greater the probability that people close to one – family and friends – will themselves have died, increasing a sense of isolation. In a world heavily focused on work outside the home, at least since the nineteenth-century industrial revolution, retirement, especially for men, has been for many a dramatic transition affecting one's sense of self-worth (does writing books such as this stave off the otherwise inevitable?). Ageing is not for wimps, as an American friend once put it, stoically facing up to some of his own problems. It is not surprising therefore to find an increase in the incidence of depression and anxiety diagnoses in old age. At the very least, there is likely to be a withdrawal into oneself. The very reduction in the brain's sensitivity and responses to external stimuli is likely to make the private life of the mind more important, as one calls on memory to help take stock of the life one has lived. How much this will change over the next decades with the dramatic greying of the population remains to be seen. What is clear is that both the bio- and the social dimensions of the latter end of the life cycle are very different for

people now in their seventies and eighties from those for the generation that reached that age only half a century ago.

Leonard Hayflick, one of the pioneer theorists of ageing, points out[4]
that as in the wild most animals succumb to predators, accident or
disease well before reaching their potential maximal life span, ageing
scarcely rates; the smallest physiological decrement, for instance in speed
of running, for predator or predated, increases vulnerability, so the oldest,
like the youngest, are most at risk. It is only with humans and protected
domesticated or zoo animals that ageing becomes apparent. There is no
specific mechanism of ageing, nor one specific cause, and, although
biological ageing follows a reasonably predictable pattern, the process
does not map simply on to chronological age.

Amongst genetic factors the most striking is a rare genetic disorder
in humans – progeria – which results in premature ageing and senility,
and those with the disorder tend to die, aged, in their twenties or even
earlier. Many other gene differences may have subtle effects on the ageing
process, but all are to a greater or lesser extent context-dependent. Few
such contextual effects in humans can be as dramatic as the differences
between the queen bee and her genetically identical worker bees. Queen
bees can live for as long as six years, whilst workers do not survive
beyond six months, the difference being due not to genes but to whether
or not they are fed royal jelly during the larval stage.

Ageing is emphatically not a disease, nor is it one specific process,
any more than is early development; it is an all-embracing term for a
long-drawn-out phase of the life cycle that in one sense begins at birth.
Rather, the term subsumes an accumulation of factors, genetic and developmental, that together and progressively reduce physiological and
cognitive efficiency at different rates in different people. The date that
ageing is considered to start is thus purely arbitrary; depending on taste
you could choose that point at which maximal neuronal cell numbers
have been achieved, or that of maximum fertility, for instance in men
when the sperm count is highest. Ageing affects the organism as a whole
and, within each creature, all body tissues; different cell and tissue types
age differently.

In general, cellular and tissue ageing reflects a breakdown in cellular
homeodynamics and regulation. Most cells in the body have a finite life,
after which they are renewed from progenitor cells. It was believed for
quite a long time that death of cells in ageing was genetically

programmed, rather as is apoptosis during development. However, in the 1980s, Hayflick provided evidence that, rather than a specific program, there was a limit to the number of cell divisions that could occur, related to changes in chromosome structure. However, there are problems even for those cells that are renewed. Our bodies are subject to a continual battering from the environment. Even before the modern chemically perfused and polluted age, cosmic radiation has always been present. So have so-called 'free radicals' – highly reactive groups of atoms carrying an unpaired electron such as so-called 'reactive oxygen' or 'superoxide'. Free radicals are generated during respiration, and can react with DNA and proteins. With radiation and free radicals comes the inevitability of tiny mutations in cellular DNA (both nuclear and mitochondrial), the odd replacement of one nucleotide letter by another, or a deletion or an addition, and these errors are then more or less faithfully copied into the next generation of cells. So minor mutations accumulate, resulting in incorrect reading of the DNA code and hence malfunctioning proteins. For example, damage to mitochondrial DNA can affect the efficiency of cellular energy generation. Although cells do possess DNA repair mechanisms in the form of enzymes that can snip out the offending nucleotides and stitch in more appropriate replacements, it is a battle that is ultimately bound to be lost. There are, after all, only so many times one can darn a sock.* Interestingly, there are mutations that affect the ability to eliminate free radicals, and there is some evidence that species with more efficient protective mechanisms for doing so have an enhanced life span.

Furthermore, whilst natural selection mechanisms will inevitably weed out gene variants or combinations which are so deleterious that the person who possesses them cannot reproduce, there are no such pressures to select against deleterious genes that are expressed only later in life, when people are beyond reproductive age, as they will already have passed on these variants to their offspring. On strictly interpreted ultra-Darwinian lines, there is thus no evolutionary pay-off to selecting out genetic variants such as those responsible for Huntington's disease with its relatively late onset, typically post-childbearing. The same is true for

---

* This, like other comments in this chapter, reflects my own historical and geographical location. I'm not sure whether in today's throwaway culture socks do get darned any more, as they were in my childhood.

Alzheimer's and Parkinson's diseases where the genetic risk factors that contribute to them only manifest themselves phenotypically later in life. It is interesting that these are specifically human diseases; there are no 'naturally' occurring animal equivalents, although genetic manipulation can now produce them in the laboratory. There is (as usual) an equally Darwinian counter-argument for the merits of longevity in the human case. As infants require a long period of care before they can survive independently, there is an advantage to having an older generation of post-reproductive grandparents around who can impart experience and share in childcare. This argument has been used to explain why human females, unlike those of other species, experience menopause, leaving them freer to care for their grandchildren. (However, this says nothing about the noticeable tendency of elderly and usually wealthy men to father children with much younger women, rather than take their share of grandchild rearing.)

There has been a steady increase in life expectancy in industrial societies over the last couple of centuries, calculated as the chance, at birth, of living to a given age. Much of the increase is accounted for by the decrease in perinatal mortality, thus reducing the number of early deaths that lower the average expectancy. However, even if one discounts this by considering only the probable number of years of life left once one has reached, say, sixty-five, there has been a steady increase, resulting in a substantial shift in the age profile of the population – the so-called 'greying of nations'. The exception that illustrates most dramatically the relevance of living conditions to length lived is Russia, where, following the collapse of communism from the 1980s on, there has been an actual decline in life expectancy.

A good deal of the improvement, setting aside the social catastrophe out of which Russia is only slowly struggling, must be due to improved nutrition and living conditions, as well as better healthcare in the industrialised nations. (I am well aware that much of this improvement has been achieved on the backs of the poorer, developing and third-world countries, where perinatal mortality is much higher and life expectancies much reduced, not least as a consequence of the AIDS epidemic.) The current spread of obesity in industrialised countries may well slow or even reverse this trend to longer life as, at least in laboratory rodents, mild calorie restriction is one form of environmental manipulation that has long been known to prolong life. However, the factors contributing to longevity are certainly complex; some populations seem to have a longer average life expectancy than others, the Japanese being a prime example. How much a fish-eating diet is responsible for Japanese longevity, or, allegedly, yoghurt for that of centenarians from the Caucasus, remains a matter of speculation. Meantime international agencies such as the World Health Organisation have responded to the global pattern by initiating 'healthy ageing' programmes, under the slogan of adding 'years to life and life to years'.

The likelihood of surviving longer today than ever before is true for both sexes, but women live on average three or four years longer than do men. The reasons for this are widely debated but obscure. Biologically based explanations have ranged around women's possession of two rather than one X chromosome to, perhaps inevitably, men's higher testosterone levels. In any event, the secular increase in life expectancy for both sexes has done little to diminish the discrepancy between them. Boys and men die off faster than women at all ages, from infancy onwards. The consequence is that there are proportionately many more older women amongst the very elderly.

The ageing brain

Within the general context of ageing, there are many ways in which the brain is particularly sensitive. The fact that neurons do not divide increases their vulnerability to the type of error catastrophe that comes from attack by radiation or free radicals. It was for a long time part of conventional wisdom that because neurons did not divide, there was a steady loss in neuronal number with age. This turns out to be one of biology's many urban myths, like the often repeated claim that 'we only use ten per cent of our brains'. It is very hard to track down the origin of such claims, and, whilst the latter is unlikely to be heard in neuro-scientific circles, the former has found its way into standard textbooks.[*] But, as I mentioned previously, recent research has shown that the brain retains a small stock of stem cells from which neuronal regeneration is

---

[*] Including, as it happens, my own *The Conscious Brain*, written in the 1970s, where I made a calculation of its implications that I now see is wholly erroneous.

indeed possible, whilst cell loss is patchy – at least in normal ageing. What does happen is that the brain shrinks with age – on average amongst European populations by as much as fifteen per cent between the ages of fifty and sixty-five. Much of this loss is due to cells themselves shrinking as they lose water, and ventricles and sulci enlarge, although there is also some neuronal loss in particular brain regions. The extent of cell death is affected by hormonal processes. Chronic stress, for instance, increases cortisol production, and chronically increased cortisol in turn accelerates cell death in regions of the brain like the hippocampus whose neurons carry receptors for the hormone. Such factors may help explain why neither the shrinkage nor the cell loss is uniform across the brain, but particularly prevalent in parts of the frontal lobes, hippocampus, cerebellum and basal ganglia.[5]

With age, there is a general diminution in blood supply to key brain regions, and, as the brain is especially dependent on its rich supply of blood through the circulatory system, both to provide glucose and oxygen and for the removal of waste material, such a diminution may be a factor in cell death, especially in the cortex. More drastic interruption to the blood supply, through for instance the furring up of the arteries that goes with high-fat diets and high blood cholesterol levels, diminishes the supply further and hence increases the chance of a stroke – the temporary shutting off of glucose and oxygen to particular brain regions which results in neuronal death.

Not only is the brain dependent on the nutrient supply from the blood; ageing carries with it changes in the immune system and hormone levels, including both cortisol and also steroids such as oestrogen and testosterone. Such fluctuations will affect their interaction with neuronal receptors in the brain, notably the hippocampus. Also, of course, there are changes in sensory inputs. The lens of the eye thickens, reducing the light falling on the retinal cells, the auditory system loses its capacity to detect higher frequency sounds; changes in the neuronal maps relating to inputs from other body regions reflect patterns of use and disuse. Changes in body physiology – sensitivity to heat and cold and the ability to thermoregulate, changes in the capacity of the digestive system, changes in body image with decline in muscular strength, loss of sexual appetite or ability – all alter perception and consciousness.

There are characteristic biochemical changes with neuronal ageing that are clearly deleterious. A number of metabolic processes slow down, notably the rate of new protein synthesis. There are conspicuous changes in the structure of the neurons; the fine fibres – microtubules and neuro-filaments – that run along the length of axons and dendrites tend to become tangled as changes occur in the chemistry of one of their key protein components, tau. Deposits of insoluble yellowish granules, complexes of proteins and sugar molecules called advanced glycation end-products (AGEs) accumulate. Meanwhile, in the space between cells, plaques of insoluble protein – amyloid, derived from the breakdown of a protein that spans the neuronal cell membrane, the amyloid precursor protein – develop. Neurotransmitter levels decrease. Physiologically, there are subtle changes in synaptic transmission and a shift in the waveform of the average evoked response as detected by EEG.[6]

However, it is important to emphasise that to speak of change does not necessarily carry the unspoken corollary of 'for the worse'. Although loss of neurons is generally regarded as one of the negative features of ageing, the space that the neuron occupied may become filled with glial cells or dendritic and axonal branches from adjacent cells, at least during early senescence, thus perhaps actually enhancing connectivity and complexity amongst the remaining neurons – and as, increasingly, astro-cytes are seen as playing an important role in brain function, swapping the odd neuron for an astrocyte may have unanticipated benefits. This type of cell death, as opposed to that caused by strokes, may therefore be seen almost as an extension of the apoptosis that is a crucial feature of early development. In later ageing, however, dendritic branches atrophy once more, and synapses in many key regions, notably again the hippocampus, tend to shrink and die back.

Although most functional changes of any significance are profoundly and inextricably mediated by social and physical context, some at least are pretty much context-independent. Take for example the eye-blink reflex, discussed in the last chapter in the context of cerebellar processes. As I mentioned, both rabbits and humans can be taught to associate a sound or light flash with the arrival of a puff of air to the eyes a few seconds later, and after a few pairings of signal and air-puff, will invol-untarily blink at the signal. The numbers of trials required before the reflex is established increases with age.[7] We are thus it seems, slower to learn even a basic reflex such as this as we get older. Speed of acquisi-tion of the eye-blink reflex has become a frequently used test for drugs which have the potential of enhancing learning and memory in older people – a topic to which I will return. Once learned, however, the reflex response is as well established in older as in younger people and is thus independent of age.

Other skills and abilities also change – thus there is an increase in the range and complexity of language with age, but also an increased frequency of mistakes, forgetting words or misnaming objects. In general, ageing may be associated with a slowing in ability to learn new material, a loss of adaptability to new contexts (certainly true for me), but also with improved strategies for remembering once-learned skills and abilities (which I hope to be also true for me, but for which I cannot vouch!). This cognitive ageing is not all-or-none.[8] Thus one of the common complaints as people age is of deteriorating memory. Yet some types of memory – such as memorising digit or letter spans – scarcely decline, a typical result being a drop from an average of 6.7 items for people in their twenties to 5.4 for people in their seventies.[9] Elderly people recall fewer items, but there is no age difference in the rate of forgetting. By contrast, working memory for reading and interpreting texts or making mental calculations does show a decline with age, perhaps because of a reduction in processing speed.[10] Episodic (auto-biographical) memory is more vulnerable than semantic memory; procedural memory (remembering how to do something) is the most robust. Interestingly, variability between individuals increases with age; the difference between the best- and the least-well-performing people on these cognitive tasks becomes greater, presumably reflecting differences in life experiences affecting the rates of ageing.

There are two ways of looking at this. One is that it shows the truth of the adage about the difficulty of us old dogs learning sophisticated new tricks. The second is that slowness in decision-making allows time for better decisions to be made. This is what used to be called wisdom, in a period before the current worship of youth and speed, and must have contributed towards the greater respect said to have been paid to older people in pre-modern societies. Especially in pre-literate cultures, the elderly – however defined, as life expectations were lower – were the repositories of experience and knowledge that needed to be passed on to subsequent generations. Modernity, with its seemingly inexorable pressure for speed-up, and its rapidity of innovation, has changed much of this.

I argued in Chapter 5 that we learned at an early age how to extract regularities from our environment, and thus also learned – a sort of meta-skill, if you like – *how* to learn and remember. This strategy must have been tremendously important in contributing to success in the early period of human evolution and migration into differing environments, when it was unlikely that the regularities learned in childhood would change greatly over a lifetime. However, in the modern industrial world, rates of techno-logical and social change are so fast that one generation's experience, accu-mulated knowledge and wisdom are of diminished value to their successors, helping perhaps to contribute to the sense of disorientation often felt by older people.*

What does become clear is why such phenomena of biological ageing need to take account of the changing social context within which each person's life cycle – and hence the private life of the mind – is embedded. Researchers are increasingly aware of the problem of interpreting the results of cross-sectional studies – that is of a group of people now all about the same age, whether in their forties or their seventies. Today's seventy-year-olds have as a group very different life histories from those who will be seventy in thirty years time. To better interpret the changes that go with ageing it is important to make longitudinal studies – of the same group of people over many decades. This is also why it has become fashionable amongst health workers to encourage reminiscence amongst older people, to encourage recall and hence the emotional and cogni-tive memory requirements that go with it, but in a more holistic sense to help each of us as we age to both take stock of and make sense of our life's experiences.

Neurodegeneration – Parkinson's and Alzheimer's diseases

The centrality of memory to our sense of personal individuality, to provide coherence to our life's trajectory, is why the diseases that rob people of their memories or personhood – the dementias – are so devas-tating both for those with the disease and those who love and care for

---

* As an older person myself I am increasingly aware of this. Few of the labora-tory techniques I learned in my student days are of any relevance to today's science, and much of the lab equipment of my youth can now only be found in science museums.

them. With increasing longevity, the incidence of these diseases also increases. Although there are many types of brain insults that can render people in varying degrees mentally damaged or incapable, from head injuries to stroke, the most common are a variety of neurodegenerative diseases in which progressively neurons die, neurotransmission is reduced or malfunctions and the person's capacities, mental and/or physical, steadily deteriorate. The best known of these are probably Parkinson's and Alzheimer's diseases, named for their respective discoverers. Although both have known genetic and environmental risk factors, for both the best predictor is age; the older a person is the more likely she is to begin developing the disease. Parkinson's now affects perhaps 100,000 people in the UK, Alzheimer's 800,000, with both set to rise over the next decades as the population ages. By the age of 80 one in five people are said to suffer from Alzheimer's (AD) and some epidemiologists, projecting these figures forward, have claimed that if humans were to reach their supposed maximal age, all would have the disease. Others, however, vigorously contest the claim, insisting that AD is a specific disease, not an inevitable corollary of the ageing process.

Both the signs and the neurological lesions of Parkinson's are somewhat easier to interpret than Alzheimer's. Early on – symptoms typically begin appearing at around age sixty – affected individuals show rhythmic tremors in hand or foot, especially when at rest. Later they become slower and stiffer, and have problems balancing; still later they may become depressed and lose mental capacity. Although the causes that precipitate the disease are not well understood (although free radicals have been implicated as amongst the culprits), the neural mechanisms that are affected are pretty clear. In particular, neurons in the substantia nigra, a region within the brain stem whose connections run via the thalamus to the motor cortex, die. These neurons are the synthesisers of the neurotransmitter dopamine. Whereas in normal ageing some 4 per cent of dopamine-producing neurons die during each decade of adulthood, in Parkinson's 70 per cent or more are lost.[11] The discovery led in the 1960s to the development of l-dopa, a precursor chemical to dopamine, as a treatment for the disease. However, although it was once hailed as a miracle drug, its effects are limited because tolerance builds up quickly, and there are a range of bizarre adverse reactions, including schizophrenia-like symptoms, dramatically brought to life by Oliver Sacks in his now classic book *Awakenings*.[12]

More recent attempts to treat Parkinson's Disease have focused on the possibility of actually replacing the dead and dying dopamine-producing cells by brain transplants. An initial claim by a Mexican research group that the symptoms of Parkinson's could be reversed at least transiently by injecting cells derived from human foetal tissue into the patient's brain was soon discounted, but intense research from the 1980s on in Sweden, the US and UK began to produce more optimistic accounts. Embryonic or foetal cells injected into the brain did appear to survive, to put out dendrites and axons, and make what at least seem like functional synaptic contacts with neighbouring neurons, but longer-term survival of the grafts remained uncertain, and it wasn't clear if they were actually replacing the lost tissue or simply acting as local dopamine mini-pumps. With the development of more sophisticated genetic engineering techniques in the 1990s, attention shifted to the possibility of restoring function to the endogenous neurons by injecting transformed carrier viruses containing DNA for the necessary dopaminergic enzyme systems. With the new century, attention has shifted once more to the possibility of using pluripotent stem cells derived from human embryos. The ethics of such use has been sharply contested and is still illegal in many countries; in Britain, permissive legislation has been pushed through Parliament as part of the government's agenda of making Britain an attractive country for biotechnological research, despite the ethical implications.[13] Whether such methods, even if permitted, will turn out to be effective is still an open question, though, because of the specificity of the brain lesions involved, Parkinson's is the most likely condition to benefit from such stem-cell technology. Here it differs from Alzheimer's disease. Alzheimer's is often linked with Parkinson's as a potential candidate for stem-cell treatment but, in my view, the more diffuse nature of the brain lesions and the very different biochemistry involved make it much less promising.

The early symptoms of Alzheimer's include loss of episodic memory for recent events – such seemingly trivial matters as 'Where did I leave my car keys?' and 'When did I go shopping?' Memory loss steadily gets worse, for names and places which should be familiar, coupled, scarcely surprisingly, with confusion, depression, anxiety and anger – delusions about people stealing one's property are not uncommon. The progression of the disease eventually requires permanent hospital care, as a progressive but often prolonged deterioration occurs. The long decline of Ronald Reagan, diagnosed with incipient Alzheimer's even during his presidency in the 1980s, but who survived until 2004, is evidence of just how long life can be preserved, although whether it remains a life worth living is debatable. It is the loss of memory, or the ability to recognise even the sufferer's own husband, wife or children that makes the disease so desperate, not only for the person affected but for those close to him or her as carers. This pattern of decline was once simply classified as senile dementia; the specific condition known for its discoverer Alois Alzheimer was regarded as a relatively rare form; today it is seen as the most common cause of dementia, with other variants being lumped together as 'senile dementia of the Alzheimer type' or SDAT.

Brain shrinkage, cell death and damage occur in all forms of dementia. What Alzheimer discovered in post-mortem examinations was something more. The dead and dying neurons were full of tangled fibrils and the spaces between the cells contained characteristic plaques of an insoluble material christened amyloid, for its resemblance to starch granules* (Fig. 7.1). Unlike the Parkinson case, where cell death is confined to a particular class, in AD neurons across the brain are affected, but amongst them the earliest to die are in the hippocampus. As the hippocampus is intimately involved in the early stages of memory formation, it is therefore not surprising that memory loss is one of the early features of the disease. The only unambiguous confirmation that a person had Alzheimer's still comes from post-mortem examination of the brain, as cognitive decline, and the shrinkage of key brain regions such as the medial temporal lobe, which can be determined by MRI scan, can have other causes (Fig. 7.2).

However, although plaques and tangles are characteristic of AD, they both occur in 'normal' brains too, gradually increasing in number with age. It is the quantitative change in their number that produces the qualitative difference between normal ageing and AD. AD is unequivocally a biochemical disorder, although its causes are complex and multiple. A rare variant, familial AD, accounting for some 5 per cent of cases, has a very specific genetic origin. Familial AD has a much earlier age of onset – typically in a person's forties – by contrast with the sporadic type. For this much more common form, a number of risk factors are known, some genetic and others environmental. Apart from age, the best predictor

---

* One of the two sugar chains of which starch is composed is called amylose.

Fig. 7.1 *Amyloid plaque.*

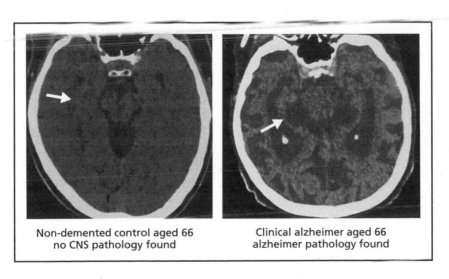

Fig. 7.2 *MRI scans of medial temporal lobe in a 'normal' subject and one suffering from Alzheimer's Disease; note the four-fold reduction in the thickness of the brain tissue at the arrows.*

of the disease is being female. Even allowing for the fact that women live longer than men, age for age, women are disproportionately at risk. One of the potent genetic risk factors is the particular form of a protein called Apolipoprotein E, which plays an important role in transporting cholesterol and other lipids across the cell membrane. There are at least four known alleles of the DNA sequence coding for ApoE; people with the ApoE 2 form are least likely to develop AD; those with the ApoE 4 form are the most likely,[14] with about a 50 per cent chance of developing Alzheimer's by the time they are seventy. Other genes, including those coding for a class of proteins disturbingly known as presenilins also contribute to risk. However, none are predictive; those with ApoE 4 may be spared the disease, those with ApoE 2 may develop it.

Evidence about the environmental risk factors is equally tenuous. Head injury and general anaesthesia whilst young are said to be risk factors. About thirty years ago there was a scare that aluminium in the diet might be involved, and many families got rid of their aluminium cooking utensils. The scare originated from the discovery that amyloid plaques have a high concentration of aluminium, but it is now clear that this is because the plaques trap the metal, scavenging it from the surrounding extracellular fluids, rather than the aluminium causing the plaques to form. Free radicals and prions – those deformed proteins involved in Creutzfeld-Jacob disease and implicated in the transfer of the disease from cattle to humans in the BSE epidemic – have been blamed, but without strong evidence. The greater prevalance in older women than in men had suggested that the loss of oestrogen postmenopause might be responsible, and there was some encouragement for this view from epidemiological studies suggesting that hormone replacement therapy offered a degree of neuroprotection.[15] However, the more recent international multicentre trials, whilst not yet complete, are said to be dampening some of the early optimism. Perhaps more interesting is that the risk of contracting Alzheimer's slightly decreases with increased educational level and a life involving mental rather than manual labour. This has generated a sort of 'use it or lose it' argument and suggestions that the best way to avoid the disease is to keep one's brain active in old age, even if only by doing crossword puzzles or knitting. This would seem sensible advice to all of us as we age, as withdrawal from active engagement in the world can precipitate a spiral of decline, but its specific relevance to AD is unclear.

Whatever the more distal causes, the immediate biochemical cascade that leads to the formation of the plaques and tangles is now relatively well understood. There are two protein players in the drama: the tau protein – a component of microtubules – which constitutes much of the tangles, and the amyloid precursor protein, APP, the breakdown of which results in the beta-amyloid peptides that compose the plaques. A battle-royal between two schools of thought as to which is the primary event has raged for several years. With the heavy schoolboy humour that characterises many lab scientists, the rival groups were named Tau-ists and B-apptists; the issue is still not settled, but the weight of evidence is on the side of the latter.[16]

What seems to happen is this. APP is a protein embedded in the neuronal cell membrane, where it is one of a family of molecules which function both in signalling between cells and also in maintaining their configuration – notably at the synapse. In the normal course of things, APP is broken down by enzymes called secretases to release fragments, one of which, called sAPP acts as a signal in processes including neuronal growth and synaptic plasticity. However, there is one version of the secre-tase enzyme which breaks off the wrong segment of APP, producing a 42-amino acid sequence, the infamous beta-amyloid, which then accu-mulates into plaques. Meantime the failure of the sAPP to do its stuff affects the internal structure of the neurons, resulting in the tau protein changing its structure and collapsing into tangles and the death of the affected neurons. Mutations in the protein sequence of APP are respon-sible for some of the familial, early onset, cases of AD. The presenilins affect which form of secretase is present and therefore whether the APP is cut in the appropriate places or whether beta-amyloid accumulates.

As I've said, it is hippocampal neurons that tend to die first in Alzheimer's disease, which presumably helps account for the early effects of the disease on memory. As one of the major neurotransmitters in the hippocampus is acetylcholine, and it is known from animal experiments that blocking acetylcholine function prevents memory formation, much of the effort to ameliorate the effects of AD until recently focused on developing drugs that prevent the destruction of the acetylcholine and hence increased its functional lifetime in the hippocampus. Three such drugs are currently licensed, although their efficacy is not great and many people develop unpleasant adverse reactions to them. A fourth drug, interacting with one type of glutamate receptor, is also now available.

These interactions, however, are all downstream of the primary lesion, the incorrect processing of APP. My own interest in the molecular processes involved in memory formation led me a few years ago to begin exploring the normal role of APP in memory in my experimental animals, young chicks, and we[17] soon discovered that if APP was prevented from functioning in any way, our birds could learn but not remember – and more importantly, I've since gone on to show that a small peptide fragment derived from sAPP can prevent this memory loss, and protect against the toxic effects of beta-amyloid. I've written about this elsewhere,[18] so I'll not go into more detail here. Whether or not the peptide turns out to be a useful agent in the efforts to ameliorate the effects of Alzheimer's disease in humans remains to be seen.

Finale

Death, like the rest of life, is a bio-social event. Whether we die having added years to life and life to years, as in the WHO campaign, or sans eyes, sans teeth, sans everything, in Shakespeare's elegiac lament on the seven ages of man, or even many years before our allotted span, is contingent. It depends, for sure, on throws of the genetic dice, but even more upon the conditions of our life, the social context that places some of us in poverty, disease and starvation, others in the relative comfort of the two-thirds society of advanced industrial nations.

I have argued throughout these seven chapters that we can understand the present only in the context of the past. Evolutionary history explains how we got to have the brains we possess today. Developmental history explains how individual persons emerge; social and cultural history provide the context which constrains and shapes that development; and individual life history shaped by culture, society and technology terminates in age and ultimately death.

CHAPTER 8

What We Know, What We Might Know and What We Can't Know

Nothing in biology makes sense except in the light of its own history

THE LAST SEVEN CHAPTERS HAVE TOLD THE STORY OF THE EMERGENCE OF human brains and minds. My perspective has been in the broadest sense historical, on the grounds that the past is the key to the present. I have offered my best understanding of what today's science has to say about how and even why brains arose in evolution, how they, and especially the human brain, develop, mature and age. The time has now come for me to try to pull together these understandings in this ostensible Decade of the Mind. For all our immense and expanding knowledge of the brain's microstructure, the subtleties of its biochemistry and physiology, what do we really know about the neuroscience of the mind? Perhaps this very concept is meaningless. What, if any, are the limits to our possible knowledge? Are there things we cannot in principle know? Can the explicit biosocial and autopoietic framework I have adopted make possible an approach to that traditional problem of Western science and philosophy, the relationship(s) of brain to mind? I realise that even within such a framework a natural scientist's understanding is bound to be inadequate. 'Solving' brain and mind in the lab isn't the same as doing so in our daily life. In the lab we can all aspire to objectivity, examining the workings of other brains – or even imaging our own – yet we go home in the evening to our subjective, autobiographical world, and aspire to make personal sense of our lives and loves.

Neuroscientists must learn to live with this contradiction. My experience of pain or anger is no less 'real' for my recognition of the hormonal and neural processes that are engaged when 'I' (whatever that 'I' may be) have that experience. Biological psychiatrists who may be convinced in their day-to-day practice that affective disorders are the results of disturbed serotonin metabolism will still discover existential despair beyond the 'merely chemical' if they sink into depression.[1] Neurophysiologists who can plot in exquisite detail the passage of nervous impulses from motor cortex to the muscles of the arm feel certain none the less that they exert 'free-will' if they 'choose' to lift their arm above their head. Even those most committed to a belief in the power of the genes somehow assume that they have the personal power to transcend genetic potency and destiny. When Steven Pinker so notoriously put it, 'if my genes don't like what I do, they can go jump in the lake',[2] or less demotically, when Richard Dawkins concluded his influential book *The Selfish Gene* by insisting that 'only we can rebel against the tyranny of our selfish replicators',[3] they merely gave naïve vent to the rich inconsistency we all experience between our 'scientific' convictions and our own lived lives.

These inconsistencies are part of the reason why natural science is incomplete. There is more than one sort of knowledge, and we all need poetry, art, music, the novel, to help us understand ourselves and those around us. This is a concession that not every biologist is prepared to make. Some, like EO Wilson, call for a 'consilience' in which these other knowledges are subordinated to those of the natural sciences.[4] Wilson is but one of those seeking to 'explain' the emergence of art, literature and music as arising out of humanity's evolutionary past, perhaps as devices for attracting mates.[5] The triviality of such an all-embracing and ultimately irrelevant 'explanation' would scarcely be worth dwelling on had it not become so much part of cultural gossip.[6] The point I wish to make here, though, is different. The problems of reconciling objectivity and subjectivity, especially in studying the working of our own brains, for sure arise in part out of the intractability – or seeming intractability – of the questions we are trying to answer. That intractability is, at least in part, a consequence of the way our thinking has been shaped and limited by the history of our own science, an issue that I have hitherto largely avoided discussing.

Indeed, despite my insistence on historicising the emergence of our modern brains, I have conspicuously failed to explain how 'we' – that is, twenty-first-century neuroscientists – acquired the knowledge that we claim. That knowledge has been influenced by the history of our several sciences, the philosophical approaches we have variously brought to our subjects of enquiry, and the developing technologies available to us. Scientists tend to write about our subjects as if blind to our own pasts, and certainly deaf to other traditions of knowledge, particularly those derived from cultures other than our own. However, it shouldn't really be acceptable. If, as I have argued, the brain cannot be understood except in a historical context, how much more so is it the case that our *understanding* of the brain cannot itself be understood except in a historical context. Our knowledges are contexted and constrained, so that both the questions and the answers that seem self-evident to us today were not so in the past and will not be in the future. Nor would we see things as we do had the sciences themselves developed in a different socio-cultural framework.

The form our biological understanding takes reflects its own ontogeny. Twenty-first-century science emerged in recognisably modern form *pari passu* with the rise of capitalism and Protestantism in north-west Europe during the seventeenth century.[7] Physics came first, Chemistry followed in the eighteenth, geology and biology by the nineteenth century. The rise of these new sciences both shared and questioned the previously ruling assumptions about the world, from the centrality of the earth to the supremacy of humans, and these assumptions helped direct the questions and answers that the sciences provided. But just suppose that science, which incorporates our view of ourselves, our place in the world, and the rules of procedure for discovering these things, had spread outwards from other centres of culture – for instance, if modern biology had its roots in Buddhist philosophy, Chinese physiology or Aryuvedic medicine. Would we then be so troubled by those wearisome dichotomies of mind and body, of nature and nurture, of process and product, which still pervade the thinking of even the most hard-nosed, reductionist neuroscientist?* Would our sciences be so

---

* There has been one entirely Western approach to science which evades this critique – that of Marxism's dialectical materialism, which insisted that complex processes and phenomena could not be collapsed into simple physicalist models. This non-reductionist approach to neuroscience and psychology

committed to the primacy of physics with its 'ultimate' particles and forces, the mathematisation and abstraction of the universe, deleted of colours, sounds, emotions . . . of life itself? Might we have, as Brian Goodwin has put it,[8] a science of qualities rather than one of quantities? Or would other questions, ones that seem to us, working within our Western tradition, to have self-evident answers, instead be as conceptually troubling?

Sadly, these are questions I find it easier to ask than to answer. There remains a silence, at least for us in the West, concerning those other sciences. Even Joseph Needham's monumental history of Chinese science[9] has little to say on the themes of mind and brain, whilst most historical accounts of Western neuroscience fall firmly into that Whig tradition which sees history as an unrelenting march of progress from the darkness of ignorance into the bright light of unassailable modernity, of what one recent writer has called our 'neurocentric' world,[10] achieved by a succession of great men. (The choice of gender here is made advisedly.) These other non-Western traditions have been lost or vulgarly subsumed within the mumbo-jumbo of new-age mysticism, with its confused conflation of words and concepts grabbed almost at random from different cultural traditions, so that immune systems and 'neurolinguistic programming' rub shoulders with Yin/Yang, cranial osteopathy and aromatherapy.* The point I am making is simple and has been amply discussed by those working in what has come to be called the social studies of science, although often disregarded or rejected by natural scientists themselves: the search for 'truth' about the material world cannot be separated from the social context in which it is conducted.

Despite their resistance to suggestions that our work is thus ideologically framed, all natural scientists will none the less concede that

---

guided research during the earlier years of the Soviet period in Russia with conspicuous success, in the hands of such figures as Vygotsky, Luria and Anokhin. The catastrophe that befell genetics in the 1940s and 50s at the hands of Stalin and his agent Lysenko was felt more mutedly in psychology and neuroscience, when in the late 1940s the 'Pavlov line' became more rigorously enforced. However, the enthusiastic adoption of Anglo-American reductionism in the former Soviet Union since the collapse of communism in the 1980s has largely obliterated that potentially fertile tradition.

* Homoeopathy, that other bogus science, with its mock nineteenth-century apothecary's chemistry is, by contrast, a Western invention.

our deployment of questions and answers, that skill which generates productive experiments, is both made possible and constrained by the technologies at our disposal. This is what the immunologist Peter Medawar once referred to when he described productive science as being 'the art of the soluble'. New tools offer new ways of thinking, new and previously impossible questions to ask. Knowledge of the brain has been transformed by successive technological advances – the anatomist's electron microscope, the biochemist's ultracentrifuge, the neurophysiologist's oscilloscope. The newest tools include both the spectacular windows into the working brain made possible by PET, fMRI and MEG, and, at an entirely other level, the ability of the new genetics to manipulate mouse genomes almost at will.

The technologies, however, constrain as well as liberate. The electron microscope fixes tissue, destroying process and dynamics. The ultracentrifuge segments cellular processes into separate compartments. The oscilloscope focuses attention on the firing of individual neurons or small ensembles, masking higher order dynamics. Brain imaging, enhanced by false colour pictures and elaborate algorithms to extract signals from noise, offers at best indications of where the action might be during any particular neural episode. Gene technology focuses on specificity to the detriment of plasticity, on single genes rather than entire genomes. Neuroscience's dependence on such technologies shapes both questions and answers. Given a hammer, everything seems more or less like a nail.

Material brains, abstract minds

Neuroscience emerges at the multiple interfaces between medicine, biology, psychology and philosophy. Its earliest rationale must have been in the attempts to treat or mitigate overt brain damage, presumably caused by head injuries. That brain injuries could result in body-function impairments was clear enough in a tradition that in the West goes at least as far back as Pharaonic times. Trepanning (Fig. 8.1) – making holes in, or cutting out sections of, the cranium – was practised by the Egyptians, and maybe goes back much earlier if the surgical marks on prehistoric skulls found widely distributed across Europe are anything to go by. The practice was not confined to the West however, but was

Fig. 8.1 *Skull trepanation. From D Diderot and JB d'Alembert,* Encyclopédie, ou Dictionnaire raisonné des Sciences, des Arts et des Métiers, *1762–77.*

seemingly widespread in the ancient world; more than 10,000 trepanned skulls have been found in Cuzco, in Peru, for instance. The holes were cut with rather basic stone tools, and, as the signs of wound healing show, their subjects were at least sometimes able to survive the operation. Just why the operations were carried out, however, is uncertain; the intentions may have been to relieve the pressures caused by injuries to the head, or to release evil spirits, or both, but the technique was in any event robustly materialist and must surely have begun to cast light on brain/body relationships. There are, for instance, Egyptian papyri describing the effects of brain injuries on motor activity and hand-eye co-ordination.

As for those crises that we now call mental illness, options included the exorcisms that predate modern psychotherapy, and a wide range of brews with pharmacological functions, of which more later. However, the role of the brain itself in mental functions remained in doubt. Egyptian and Graeco-Roman, Chinese and Aryuvedic medicine gave pre-eminence to the heart over the head, although some Greek surgeons and philosophers dissented, as in the Hippocratic tradition, which saw the brain as the body's controlling centre. Plato and Democritus divided the soul into three, associating the head with intellect, the heart with anger, fear and pride, and the gut with lust and greed.[11] For the Chinese, consciousness was associated with the spleen and the kidneys with fear, whilst the brain itself was merely a sort of marrow. Indeed, in the West as well as the East, the fluid-filled cerebral vesicles at the core of the brain seemed of more importance than the rather unpromising gooey material that surrounded them. One popular theory claimed that the role of the cerebrospinal fluid in the vesicles was to cool the blood, although Galen, in his systematisation of Mediterranean medical know-ledge in the second century CE, was able to reject this, if for no better reason than that if this were the brain's function it should surely have to be located closer to the heart. The problem is that before the advent of microscopy, it was difficult to observe much structure to the brain tissue itself, other than the divide into grey and white, the prominent blood vessels, and the strange doubling of cerebral and cerebellar hemi-spheres.

Nevertheless, our language continues to carry the history of those functional doubts. When we learn things *by heart*, we are being faithful to the older understanding that seated all such mental functions in what nowadays is regarded merely as a superbly efficient pump. And if the brain was to have any part at all in thought, it was limited to the coldly rational. Shakespeare could ask where was fancy bred, 'in the heart or in the head?', but he knew the answer that his listeners expected. It is more than four centuries since William Harvey demoted the heart from the seat of the soul to a mere pump mechanism, yet it is still symbolic *hearts* that lovers exchange on St Valentine's day, not almonds (for the amygdala), and when overcome with gratitude, we give *heartfelt* thanks. The ancient Hebrews, whose 'bowels yearned' when they were emotionally overwhelmed, were more in accord with modern understandings. At least the intestines contain neurons and the general regions of the bowels are rich with hormone-secreting organs. So today's *gut feelings* have something going for them. The metaphorical division between heart – or gut – and head remains embedded in present-day neuroscience, with its insistence on dichotomising between cognition and affect, even if today the 'seats' of both are more likely to be located a few centimetres behind the eyes.

Even if early surgeons were able to attribute some aspects of bodily control to the functioning of brain and nerves, and the loss of control to damage or disease, mental activities – thought – could be left safely to the philosophers, and its disorders to the herbalists. Aristotle classified different forms of memory, from the fading immediate present to the record of past experience and the creation of mental images without the need to relate them to body or brain parts.[12] After all, mentation seems delocalised, ungrounded in the material world of flesh and blood. St Augustine put it most clearly in his *Confessions*,[13] written 1600 years ago. How could the mind encompass vast regions of space, vast epochs of time, if simply encased within the body? How could it have abstract thoughts, envisage numbers, concepts, the idea of God? Memory was a particular mystery, a 'spacious palace, a storehouse for countless images'. But memory is capricious. Some things come spilling from the memory unwanted, whilst others are forthcoming only after a delay.[14] Memory enables one to envisage colours even in the dark, to taste in the absence of food, to hear in the absence of sound. 'All this goes on inside me in the vast cloisters of my memory.'[15] Memory also contains 'all that I have ever learnt of the liberal sciences, except what I have forgotten . . . innumerable principles and laws of numbers and dimensions . . . my feelings, not in the same way as they are present to the mind when it experiences them, but in a quite different way . . .'[16] and things too, such as false arguments, which are known not to be true. Further, he points out, when one remembers something, one can later remember that one has remembered it. No wonder that the mind seemed to soar outside the physical confines of mere brain-goo. For Augustine, unlike Emily Dickinson, it is the mind, not the brain, which is wider than the sky.

As with so much of modern biological thinking, the route to the brain as the 'organ of mind' was opened by Descartes with his insistence that non-human animals were mere mechanisms, albeit, as professedly a good Christian, open to Augustine's problems, he left the soul/mind in place for us humans. Although everything about our day-to-day functioning could be regarded as mechanical, no different from that of any other animal, humans, he insisted, can think, whereas other animals are capable only of fixed responses to their environments. Thought and soul are incorporeal entities, but could interact with the mechanism of the body by way of the pineal gland, located deep in the brain. Descartes chose the pineal for this task on two grounds. First, he believed that the other brain structures are all duplicate, in that the brain consists of two more or less symmetrical hemispheres, while the pineal is singular, unduplicated, and mental phenomena are normally unitary. And second, he claimed, the pineal is a structure found uniquely in humans and not present in other animals.

He was in fact wrong on both grounds; there are many other non-duplicated structures in the brain, and other vertebrates also possess pineal glands (with, as is now known, important functions in regulating body clocks and rhythms), but the theory-driven logic of his argument remains appealing for those who want to argue, as he did, for the uniqueness of humans: 'It is morally impossible that there should be sufficient diversity in any machine to allow it to act in all the events of life in the same way as our reason causes us to act.'[17] And he still attributed great importance to the cerebrospinal fluid, suggesting, for instance, that memories were stored in the pineal and activated by the changing pressure of the fluid bending the minute hairs on its surface in one or another direction.

Descartes' functional anatomy could trace sensory and motor nerves to and from the brain, producing the reflex actions shown in his celebrated diagrams (see Fig. 6.3). But the Cartesian diagrams end in obscurity, somewhere in the brain-goo where the soul or mind begin.[18] The obscurity was a result of the persistence of Galenic views of brain structures. Accurate brain dissections began to be performed in the mid-seventeenth century, notably in the hands of Thomas Willis during the Cromwellian period in Oxford. Willis and his colleagues demonstrated the flow of blood to and from the brain, learned to preserve the tissues in alcohol for more accurate, leisurely dissection, and turned the attention of the anatomists from the ventricles to the surrounding tissue (Fig. 8.2).[19] The stage was set for a more materialist account of the brain,

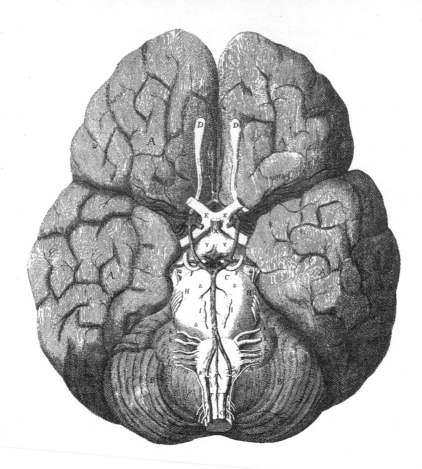

Fig. 8.2 *Christopher Wren's drawing of the human brain. From Thomas Willis,* The Anatomy of the Brain and Nerves, *1664.*

even though Willis himself simply traded the Galenic tradition in favour of a quasi-mystical alchemical one. However, as neither the Galenic nor the alchemical tradition could account in anything other than a hand-waving way for function, either at the level of nerves or of the brain as a whole, or could begin to answer Augustine's questions, the obscurity persisted for most of the following 350 years, leaving plenty of space for dualists and other mentalists to insist that the brain might enable, but could not contain the mind. After all, what was the little homunculus in the brain, which for so many years was supposed to co-ordinate all aspects of life, to which other regions supposedly reported and which in turn supervised or instructed 'lower' responses, other than a more modern version of Descartes' pineal? The rejection of the homunculus that I recount in Chapter 6 has come pretty recently in neuroscience's history; it persists in everyday speech, as the 'I,' which constitutes our concept of self.

Descartes himself must have recognised that his dualism was an uneasy and dangerous compromise, which is why he suppressed publication of much of his work during his life time, for fear of succumbing to the same fate as Galileo. The emergence of neuroscience required the wresting of control of the soul and the mind from theologians and philosophers, a battle for power that occupied the period between Descartes and the radical materialist physiologists of the nineteenth century. Willis and his contemporaries were acutely sensitive to the charge that a materialist account of the brain paved the way to atheism. This was exactly the charge thrown at (and embraced by) radical materialists such as La Mettrie by the end of the seventeenth century, when Spinoza and his followers were insisting on the unitary and deterministic nature of the universe. This pantheism was seen as akin to atheism and led to Spinoza being excommunicated by the Jewish community in Holland.

Spinoza's determinism none the less retained a view of mind as distinct from brain, operating in a sort of parallel universe.[20] This approach persisted amongst philosophers pretty much through the next two centuries. It enabled them to profess materialism and yet to ignore the brain in preference to the mind. Even eighteenth-century rationalists like Hume and Locke (who had once been Willis's student) showed a respect and reverence – almost a sense of mystery – about the mind and all its works, without feeling the need to locate its functions within the body, and their followers are still amongst us, especially in the English tradition which enables university courses to teach 'philosophy of mind' without reference to the brain at all. Other modern philosophers of mind and consciousness (Colin McGinn is an example[21]) agree that neuroscience has a lot to say about the brain, but claim that it simply cannot ever deal with such traditional matters of philosophical concern as qualia (see page 215) and consciousness. This permits a sort of unilateral declaration of independence of mind – or at least the *study* of mind – from brain. Indeed, through most of the past century even the most committedly non-dualist of philosophers have tended to be satisfied with merely rhetorical references to the brain.

Powerful psychological and psychotherapeutic schools, notably behaviourism and psychoanalysis, took the same route, at least until very recently. Freud, it is true, originally had an interest in neuroscience and neuronal theory, and in the last decade his followers have attempted to forge an alliance with sympathetic neuroscientists. For behaviourists, until their arid approach was finally killed off in the 1960s, the brain remained a black box, with inputs and outputs in terms of behaviour (for the behaviourists) or thoughts, dreams and emotions (for the analysts), but whose intimate workings could be left to the scientific equivalent of car mechanics. All that mattered was that you should know how to get into the car, or engage the mind, and drive or mentate off. Even the 'modules' with which that seemingly biological approach, evolutionary psychology, is so enamoured, are purely theoretical, with no necessary grounding in actual brain structures or functions. Although it rejoices in employing gene-speak, evolutionary psychology's genes are abstract accounting units rather than functioning DNA sequences.

It is not, of course, that anyone these days would argue that brain and mind were unconnected, but rather that their distinct discourses, their terms of reference, are so separate that no purpose is served by attempting to relate, for instance, the concept of human agency with the reductionist mechanical materialism of those who would wish to eliminate mentalese from discourse. However, such implicit dualism is increasingly unfashionable, and is unlikely to survive the continuing advance of the neuroscientists. Thus, to return to the post-Cartesian history, throughout the eighteenth century and despite the philosophers, mechanistic ideas invaded biology with increasing confidence, spurred by developments in other sciences. Sometimes indeed the advances were entirely metaphorical. Descartes' own models for how the brain worked were built on an analogy with hydraulics in which vital spirits coursed down the nerves to activate muscles, and Willis could do little better, but sometimes metaphor and mechanism fused, as when the eighteenth century saw the discovery of the role of animal electricity in the electrical discharges that caused the muscles of Galvani's dead frogs to twitch.* However, the crucial advances in understanding required other technologies.

---

* More modern metaphors, such as that of the brain as computer, also combine an ideological function – a way of thinking about the brain and its activities – with an approach to organising experiments and analysing data.

The struggles over location

Until the advent of the microscope and electrical recording and stimulating methods, analysis of brain function could only focus on the macro-
scopic, on relating function – whether sensory, like vision; motor, like
the control of movement; emotional, like anger; or cognitive, like mathematical ability – to particular brain structures. The centuries-long battle
over whether particular mental attributes and capacities are functionally associated with, emerge from, or are dependent on, specific brain
regions is a classic example of the interdigitation of limited technology
with philosophical convictions.* A convenient starting point is the early
nineteenth-century craze for phrenology: the reading of character and
capacity via the shape of one's head, a new science invented by the
Austrian anatomist Franz Joseph Gall and popularised by Johann
Spurzheim. Phrenology was based on the principle that the brain, the
organ of mind, was composed of distinct parts, each representing a
specific faculty, and that the size of these different regions, assessed
through the cranium, reflected the strength of the faculty they contained.
At first sight this makes good sense; the problems however arise firstly
in deciding how to carve up 'faculties' into distinct entities, and then
how to assign the brain structures. Phrenology's problems, to a modern
neuroscientist, begin with the former. Thus the miniature phrenology
head that sits on my own desk, created, as proudly announced on its
base, by one I.N. Fowler, has these faculties conveniently printed on to
its bald cranium: 'exploration', 'foresight', 'worship', 'trust', 'acquisi-
tiveness', 'desire for liquids' and many more, each with their own fixed
location (Fig. 8.3). As for the rationale for localising, this was apparently
based by Gall and his followers on an examination of the skulls of individuals who had in life shown particular propensities. Thus 'destruc-
tiveness' was located above the ear because there was a prominent bump
there in a student said to be 'so fond of torturing animals that he became
a surgeon'.[22]

Phrenology, despite its many celebrated adherents, was eventually
laughed out of court, but the problem of localisation continued to

---

* In its modern form the debate is not so much over anatomical location, although
that remains controversial in many cases, but over locating behavioural function to specific genes or neurotransmitters, the substance of much of the next
chapter.

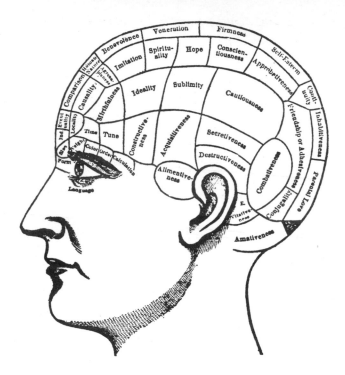

Fig. 8.3 *Phrenological map of the brain, early nineteenth century.*

perplex the leading neurologists of the day. Thus the prominent anatomist Marie-Jean-Pierre Flourens, over the twenty years from the 1820s to the 1840s, whilst accepting functional localisation in such struc- tures as the cerebellum, argued strongly that the cortex functioned as a unit, indivisible into modalities, and his view was highly influential.

The favoured approaches to resolving such questions – indeed virtu- ally the only ones possible prior to modern imaging techniques – were either to infer function from dysfunction, that is, to relate damage to specific brain areas to loss of specific abilities, or to explore the effect of stimulating specific brain regions electrically and observing the 'output' in terms of precise motor responses. Dysfunction could be achieved deliberately by creating lesions in experimental animals – routinely cats and dogs – and looking for consequent deficits. In Victorian Britain, such lesion experiments, which seemed both cruel and not very informative, served to galvanise a powerful anti-vivisection movement, the precursor of today's campaigners for 'animal rights', and, in response, the first government legislation licensing and protecting animal researchers.

The results of such experiments were confusing. Some lesions to particular regions of the cortex gave rise to specific motor or sensory deficits. However, others seemed non-specific, and instead seemed to produce deficits dependent merely on the extent of the cortex that had been removed. It was this type of observation that resulted, half a century later, in a now classic cry of despair from the memory researcher Karl Lashley, who trained rats to run a maze and then cut or removed sections of their cortex in an effort to discover 'the engram', the putative memory store. However, it proved elusive. Wherever the lesions were placed the rats still managed to move through the maze, employing hitherto unused muscles or sensory information; their deficits were, it seemed, quantitative rather than qualitative.[23]

An alternative to making lesions was to try to stimulate particular brain regions electrically, building on the increasing evidence that information was carried in the brain through electrical signalling processes. The pioneering experiments were by Gustav Fritsch and Eduard Hitzig in 1870, who, working with dogs whose brains they exposed, found distinct cortical sites which produced specific muscle responses in paws, face and neck – but on the opposite side of the body to that of the cerebral hemisphere being stimulated. This was evidence not just for localisation, but for another counter-intuitive brain property – the crossing-over that ensures that the right hemisphere relates primarily to the left side of the body and vice versa.

This crossing-over is a reflection of another aspect of the brain that caused considerable anatomical and philosophical debate. Why is the brain double, with apparently symmetrical (actually only nearly symmetrical) structures, left and right cerebral and cerebellar hemispheres? Such questions vexed philosophers as well as anatomists.[24] How, for the religious, can a double brain house a unitary soul – Descartes' problem? Or, for the more secular, as one's sense of self is normally unitary, why should there be two hemispheres, requiring an integration of activity across the corpus callosum which bridges them? Some nineteenth-century theorists insisted on the primacy of the left over the right hemisphere; thus, according to historian Anne Harrington, the criminologist Cesare Lombroso (who believed that you could diagnose criminality on the basis of facial features) saw the right hemisphere as 'evolutionarily retarded, neurotic, and generally perni-cious'.[25] The apparent enigma of the double brain persists in other, if somewhat less prejudicial, forms today. Left brain/right brain, distin-guishing polarities between masculine and feminine, cognitive and affective, verbal and visual, dates (probably) from the 1960s as part of an ongoing incorporation of partially- or mis-understood 'science-speak' into the common culture (an incorporation the terms share with casual references to immune systems, genes and DNA).

Lombroso was in some ways typical of a revival of phrenology that occurred during the late nineteenth and early twentieth century, when collections began to be made of the brains of famous men. Paul Broca himself initiated one such, and donated his own brain to it. An entire institute was built in Moscow to house Lenin's brain and the scientists appointed to examine it, and the macabre post-mortem history of Einstein's brain has been well documented.[26] Could one identify supe-rior intelligence, psychopathy or sexual deviance from differences in the macro- or microscopic appearance of dead brains? Lenin's neuroau-topsists were fascinated by the number of glial cells in his frontal lobes, to which they attributed his political abilities, but it is not so much supe-rior intellect as the rather more prurient question of sexual orientation which has most fascinated modern-day phrenologists, who claim to be able to identify differences in the thickness of the corpus callosum, or the size of certain hypothalamic nuclei between gay and straight men and women.[27]

As the localisation issue became a central problem, the obvious alter-native to animal experiments was to observe the effects of accidental damage to the brain and attempt to relate it to changes in behaviour. Inevitably such accidents are random and varied, so systematic evidence was hard to collect. Some of the most gruesome came from the battle-fields of the Franco-Prussian war of 1870, which provided a portfolio of brain injuries for enterprising surgeons to document. Indeed, wars have continued to act as a spur to neuroscience as to so many other fields of endeavour and innovation. For example, the mapping of the visual cortex and the effects on vision and perceptions of lesions to specific areas of the visual cortex was pioneered by the Japanese surgeon Tatsuji Inouye during the Russo-Japanese war of 1904, when Russian high velocity bullets caused discrete lesions to the brains of Japanese infantrymen.[28]

In the main, such patients are unnamed, or identified only by their initials (as with HM and the hippocampus, described in Chapter 6), whilst the doctors or scientists enter the pantheon of eponymity – Broca, Alzheimer and Parkinson, to name but a few. Certain individual cases have however entered neuroscience's folklore, the classic instance being the US railway worker Phineas Gage, who was injured by an explosion that drove a bolt through his skull, destroying a large region of his left orbito-frontal cortex. To everyone's amazement he lived on for several years, but with a changed temperament. From being a relatively placid, sober person, he was said to have become frequently drunk, coarse, readily angered and abusive. This might seem a not unsurprising response to such a dreadful injury, but it could also be attributed directly to the loss of the ameliorating social functions that engage frontal lobe activity. However, because Gage's change in personality was general and hard to categorise, the inauguration of the era of localisation is usually regarded as being Broca's discovery in 1861 of the very specific effects of a left inferotemporal lesion on speech in his patient 'Tan', whose implications I discussed in Chapter 4.

Fritsch and Hitzig's pioneering use of electrical stimulation was refined in the decades that followed. With the twentieth century came the development of newer electrical techniques, making it possible to implant or insert electrodes into specific brain regions and observe the effects of stimulation. The animal experiments presaged the use of such techniques on humans too, and it became standard practice to employ them in the surgical treatment of epilepsy, which involves opening the skull to locate and remove the scar tissue or glial lesions that serve as the focus for the seizures. During such operations, pioneered by Wilder Penfield in Montreal in the 1950s, it became routine to insert stimulating electrodes to discover the sites of the lesions. Doing so made it possible to record the motor, sensory or verbal responses of the patient to such stimuli. It was this technique that led to the mapping of the motor and sensory homunculi in the cortex that graces virtually every basic neuroscience textbook (Fig. 8.4).

Thus, despite Lashley's *cri de coeur*, it has become increasingly possible to relate specific brain functions – at least those involved in overt sensory or motor processes – to specific brain regions. The crucial metaphor, as it is now understood and as discussed in Chapter 6, is that of *maps*. The brain, and particularly the cortex, holds representations of the outside

(a)

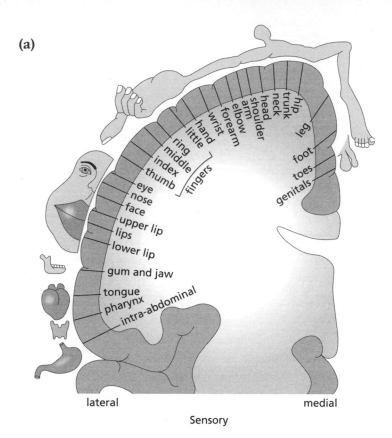

lateral   medial

Sensory

Fig 8.4 *Sensory and motor homunculi, after W Penfield.*

world, and plans of actions upon that world, in the form of multiple and complex patterns of neural activity in specific cells, in topologically defined relationships with their neighbours, which together provide models of the outside world – or at least models of how the brain and mind perceive the outside world, for ultimately what we know is not the world but our perception of it. To seek localisation of thought, emotion or action in specific areas of the cortex is, it was slowly realised, to commit a category mistake, for such processes are held not in one location but in a pattern of dynamic interactions between multiple brain regions, cortical and inferocortical. This, at any event, is how it seems to us, in the context of both the history of the brain and the history of our study of the brain.

(b)

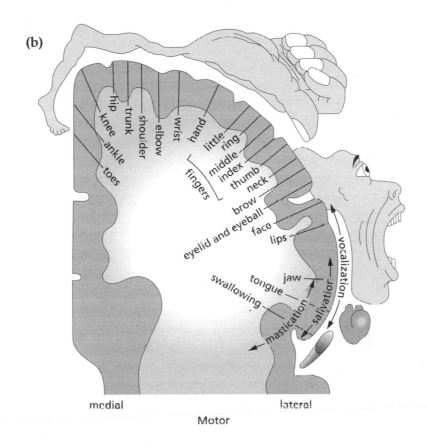

medial

lateral

Motor

Designing brains and minds

In his book *How the Mind Works*,[29] Steven Pinker argued that to help understand mental processes one might make use of the idea of reverse engineering, an approach apparently used by rival companies which involves looking at an existing designed object and asking how the engineers who created it arrived at that solution. However, in the world of biological evolution such a rationalistic approach is simply inappropriate. Asked to design a brain starting from scratch, no engineer would produce anything remotely resembling the brains we humans actually possess. However, as Chapter 2 charts, our brains weren't built from scratch.* Nor therefore, one must assume, were our minds. Evolution works via

---

* Of course it is also very rare for an engineer to design anything from scratch. Human artefacts also carry their history within their design: modern cars trace

incremental processes, always constrained by what is presently available and functioning. There is an arrow of irreversibility, in the sense that there is no turning back to some structure or function less well adapted to ensuring survival and reproduction in whatever the currently existing environmental circumstances. It is not permitted to reach a better design by a route that requires disassembling the present one if such disassembly reduces fitness. The solution may not be the best of all conceivable ones, but it is at least potentially the best of all possible and available ones. It is in this sense that evolution is a Panglossian process.

As I spelled out in Chapter 2, human brains reveal their ancestry. Their basic biochemistry was essentially fixed at the dawn of evolutionary time, with the emergence of cells. Their basic neurophysiology, with its membrane potentials and signalling by way of chemical messengers, predates the appearance of nervous systems. Neurons were invented aeons before they were assembled into brains. The puzzle of contralateralism, requiring nerves to cross over, connecting right body to left brain and vice versa, appears even in early vertebrates and their successors have had to live with it ever since. New design solutions do occasionally appear, as in the inversion of white and grey matter, which solved the neuronal packing problem by putting the cells on the surface to form the cortex, and tucking their nervous inputs and outputs within, insulated in their myelin sheaths. However, the cumbersome multilayered cerebral regions, with their seemingly overlapping or redundant functions, such as thalamus and cortex, track the steady dominance of forebrain over midbrain during vertebrate evolution, in which structures are rarely abandoned, merely overlaid by and subordinated to others. A camel, it has been said, is a horse designed by a committee. A visitor from Mars, contemplating this ramshackle brain-building with its inconvenient access, rickety staircases, odd extensions – and amazed that it worked so harmoniously at all – might draw a similar parallel.

And is it on this camel-like evolutionary idiosyncracy on which our minds, humanity's glory, must depend? Ask an engineer to design a mind, which was Pinker's real concern, and the first question would be,

---

their design roots to horse-drawn coaches; Macintosh OSX to OS9. It is too difficult, too expensive, and often not within the reach of imagination, to start with a genuine blank sheet of paper. In this sense there is an irreversibility too about the developmental paths taken by human technologies.

well, what are minds *for*? If the answer is that they are for solving complex logical problems, performing rapid computations, playing chess, storing data for subsequent retrieval and manipulation, what might emerge is something not so unlike a computer – but, as the earlier chapters have argued, this isn't only, or even primarily, what human minds do. Indeed, we do them less well than computers, which have only to deal with information. Once again: to understand present-day minds, we have to understand the evolutionary pressures and constraints that fashioned them. Our minds have all the capacities wondered at by St Augustine; they can contain the world, as Emily Dickinson should have written. We have/are selves, we are conscious, we have feelings. We love and hate, we can theorise the universe and construct philosophies and ethical values, we can invent and dis-invent gods. Above all, we are social beings, and, as I have continued to emphasise throughout this book, our minds work with meaning, not information. In some way the expansion of mental capacities, from our unicellular ancestors to *Homo sapiens*, has taken place *pari passu* with the evolution of – what? Not just the brain, I have argued, but the brain in the body and both in society, culture and history. Could a Pinkerian technologist reverse-engineer such a mind? Frankly, I doubt it. But we must work with what we have available. Is what we now know, and what we are likely to discover, about the brain enough? How far might we be able to go in relating mind to brain, or, to be less mechanistic and more dynamic, in relating mental processes and dynamics to neural processes?

Brain memory and mind memory

Back in Chapter 6, I explored this question by asking at what level, biochemical, cellular or systems, one should look to explain how the brain enables mental capacities. To approach the question, I have argued, the study of memory is an ideal case. My researching life, as I have written about it in *The Making of Memory*, has been built upon the claim that learning and memory can form a sort of Rosetta Stone, with brain language and its matching mind language inscribed in parallel on a single block, making it possible to decode, to learn the translation rules between the two. This is because, in designing experiments, it is usually easier to measure change rather than stasis. Animals can be trained on specific tasks, or humans asked to learn specific things such as word lists, the rules of chess, how to ride a bicycle – or to remember some past experience, such as what they had for dinner last Wednesday – and we can then ask what has changed, molecularly or structurally, in their brains and bodies as a result of that learning or remembering. This is how I, along with many of the current generation of neuroscientists, have endeavoured to hunt down Lashley's engram.

The Rosetta Stone metaphor seems at first sight a good one, but problems confront you as soon as you try to unpack it. The first is that even within the brain and cognitive sciences themselves there is little agreement as to what constitutes memory and how to study it. Is there even such a thing as 'a memory'? The tradition in experimental psychology, which regards the brain as a black box, and seeks merely to define abstract universal learning rules, doesn't help. Computer people and neural modellers seem to have no problem in defining memory in terms of bits of stored information – bits in the technical, information-science sense. One Oxford neuroscientist, whom I discussed in my memory book, even calculated that the hippocampus can contain some 36,000 distinct memories. But I have the utmost difficulty in deciding what constitutes 'a' memory. How might one decompose an experience, a face, the taste of a meal, the concerns of St Augustine, into such bits. For that matter how many bits are involved in remembering how to ride a bicycle? Real life, as opposed to the aridities of the learning and remembering of word lists in the clinical environment of the psychology lab, seems much too rich to delimit like this.

For sure, the taxonomy of declarative and procedural, episodic and semantic, makes a sort of sense. The progressive loss of episodic followed by semantic followed ultimately by procedural memory that occurs in Alzheimer's disease suggests that these are distinctions which reflect either different biochemical and physiological mechanisms or functionally distinct structures, or of course both, but this is not the only possible taxonomy of memory. Consider the difference between recognition and recall. Asked to describe the appearance of an acquaintance one has not seen for some time, most of us would have some difficulty – but see or hear her, and she is immediately recognised. There's a well-known children's game (Kim's game) in which the players are briefly shown a tray holding perhaps twenty objects: matches, a knife, a key, etcetera. The tray is removed and the task is then to list – to recall – all the items on it. Most adults can manage around twelve to fifteen – children can often do a bit better. So recall memory seems finite, even quite limited. However, in a famous experiment back in the 1950s, the Canadian psychologist Lionel Standing showed volunteers a series of slides of pictures or words, each visible only for a few seconds. Two days later the subjects were presented with matched pairs, side-by-side, each consisting of one previously shown slide and one novel one. Their task was simply to point to which of each of the pairs they had seen previously. To Standing's surprise, they were able to make almost error-free choices up to the astonishing number of ten thousand, after which presumably the experimenter got as bored as the volunteers.[30] Even more surprisingly, the error rate did not increase with the number of slides shown. So, unlike recall memory, recognition memory seems essentially infinite, unbounded. How can this be? No one knows.

Then there is the temporal dimension. We speak of short- and long-term memory – short-term lasting minutes to hours, and long-term lasting seemingly indefinitely – and of the transitions between them that we can study at the molecular level. (From our own animal experiments, the crucial transition occurs at around four to six hours.) Long-term memories are believed to be permanent. If this is the case, then forgetting is not erasure, but some temporary inability to access the stored memory. Indeed this seems to be what happens. I earlier described an experiment in which animals (usually rabbits) are trained to make an association between a tone (in Pavlovian terms, a conditioning stimulus) and a puff of air to their eye given a few seconds later (the unconditioned stimulus). The rabbits quickly learn to blink when the tone sounds, and before the air-puff. However, if the tone is then repeatedly presented without the air-puff, the animals eventually cease to blink to it. This is called extinction, a sort of unlearning or forgetting. There has been a long debate as to whether this unlearning is an active or a passive process, and it is now generally regarded not so much as forgetting, but actively learning not to do something previously learned. But it turns out that if, once extinction has occurred, the rabbits are given a single trial of tone paired with air-puff, they almost instantly relearn or re-remember, the association. So, in some way, despite the extinction, memory for the association must persist to be called into play the moment it is required. This means that the memory is present, even when the animal does not respond in the way that would be expected if it still remembered. How does this happen? Nobody knows.

All this means that there is a separation between 'a memory' and remembering it. So there must be some form of scanning mechanism in the brain/mind that enables memories to be retrieved. We humans can be taught how to employ this scanning mechanism more effectively, by adopting methods used in the regular memory Olympiads in which competitors perform such seemingly useless feats as remembering the entire sequence of a shuffled pack of cards after a single rapid viewing. 'Learning to remember' is apparently helped by 'placing' the things one wants to recall in sequence in a room or house through which one can later walk in one's imagination. Such mental strategies are enshrined in thousands of years of advice to aspirants wishing to improve their memory, from ancient Greek days to the present time.[31] In practice most of us, unable to recall a person's name, will spontaneously try various scanning devices, such as the letter it starts with, what the name might rhyme with, the last time we saw the person, what they were wearing, and so forth. But what are the brain correlates of such mental scanning processes? What 'scans' in the brain, and how? No one knows.

There are more complex temporalities. People who have had head injuries, or have been treated with electro-convulsive therapy, also show 'spotty' losses of memory, with older memories seemingly more vulnerable than newer ones.[32] Head injuries often result in both semantic and episodic amnesia stretching back weeks, months or years, but as the injured person slowly recovers, so memory gradually returns, except for a period of minutes or hours prior to the accident, which seems permanently lost.* The permanent loss makes sense, as it accords with the short-term/long-term transition, but why the temporal gradient of recovery? No one knows.

Meanwhile, other psychologists offer a different taxonomy and temporal timescale, speaking of working versus reference memory. Working memory is what is called into play when one is actively engaged in the work of recall, and may last only for seconds; it is the memory of the ever-disappearing present as it fades into the past. Imaging studies

---

* Beautifully described in the moving film *Who am I?* made by the TV journalist Sheena McDonald, which presents her own exploration of her recovery following horrific frontal lobe injuries occasioned when she was hit by a speeding police car. The film was shown on BBC4 in January 2004.

show that working memory involves a dynamic flux across many cortical regions. Reference memory is the hypothetical 'store' of semantic and episodic data, perhaps located in the left inferotemporal cortex, a region which imaging studies show also becomes active when memory is called upon, as in our MEG shopping experiments. How do these temporal scales and regional distributions map on to those found in the cellular and molecular studies? No one knows.

Down amongst the molecules, things seem easier, more precise. As Chapter 6 discusses, biochemical and pharmacological experiments prove that when an animal learns a new task, a cascade of molecular processes occurs that results in altered electrical properties and synaptic connections in specific brain regions, as predicted by Donald Hebb. Blocking these processes in experimental animals with specific drugs or inhibitors prevents memory being formed, or results in amnesia once it is formed, whilst enhancing them can strengthen otherwise weak memories. Some at least of these findings can be replicated in humans. Physiological studies show that the hippocampus and amygdala have important roles to play in the registering of new information and its cognitive and affective significance, so that hippocampal damage produces profound deficits in a person's ability to retain new information in longer-term memory. Whether a memory is registered or not also depends on other processes outside the brain such as stress and associated levels of circulating hormones, which interact with both amygdala and hippocampus. However, despite this apparent evidence in support of the Hebbian prediction, memories do not seem to be encoded within these specific changes in connectivity, but rather, over time, other brain regions and connections become involved. The imaging studies add to these problems of interpretation by revealing the dynamic nature of the cortical processes involved in both learning and recalling, in which even primary sensory areas such as the visual cortex seem to be involved. As to how the biochemistry and physiology can be matched to the psychological taxonomies – no one knows.

The sad fact is that even within the neurosciences, we don't all speak the same language, and we cannot yet bridge the gap between the multiple levels of analysis and explanation, of the different discourses of psychologists and neurophysiologists. As an indication of just how bad it is, one only has to look at the review papers and books produced within these two broad traditions. I chose two of the best known, each by distinguished researchers in memory. Yadin Dudai is a molecular neuroscientist, whose *Neurobiology of Learning and Memory*[33] remains one of the best in the field. Alan Baddeley is a leading cognitive psychologist, developer of the theories of working and reference memory discussed above, whose book on human memory[34] is perhaps the key text in the area. Dudai's book is full of molecules and neural circuits, Baddeley's is replete with terms like 'central executive', 'visual sketchpads' and an entirely other set of wiring diagrams of the logical interconnections between these hypothetical entities. It is hard to believe the two authors are studying the same phenomenon, and the list of references at the end of each is virtually non-overlapping.

So, despite decades of theoretical and experimental work, the application of the entire contemporary range of neuroscientific and psychological tools, from genetic manipulation via pharmacology to imaging and modelling, we don't know how memories are made, how and in what form they are stored in the brain (if 'storage' as in a filing cabinet or computer memory is even an appropriate way of speaking) or the processes by which they are retrieved. We are unsure whether memory capacity is finite or bounded, whether we forget, or simply cannot access, old memories, nor even how our most certain recollections become transformed over time. We don't know how to bridge, still less how to integrate, the different levels of analysis and discourses of the differing sciences. A single word 'memory' means different things even within the constraints of the sciences. So, it is no wonder that my Rosetta Stone metaphor of translation remains a long way from realisation, even within this relatively narrow compass. If we can't relate what is going on down among the molecules to what might comprise the workings of a visual sketchpad, we are still in some trouble invoking either to account for what happens in my mind/brain when I try to remember what I had for dinner last night. I don't want to suggest that there are never going to be answers to these questions, but simply to show how far we have still to go.

The next big step?

Ask neuroscientists what they see as the next big step in their science, and the answers you will get will vary enormously. Molecular neuroscientists are likely to offer proteomics – the identification of all of the hundred thousand or so different proteins expressed at one time or another in one or other set of neurons. Developmental neuroscientists are likely to focus on a better understanding of the life history of neurons, the forces that shape their destiny, control their migration, determine which transmitters they engage with and how experience may modify structure, connectivity and physiological function. Imagers might hope for better resolution of mass neural activity in time and space – a sort of merger of fMRI and MEG. But I suspect that underlying all of these specific advances there is something more fundamental required. We simply do not yet know how to move between the levels of analysis and understanding given by the different brain discourses and their associated techniques. Thus imaging provides a mapping – a cartography – of which masses of neurons are active under particular circumstances. It is, to use Hilary Rose's memorable phrase, a sort of internal phrenology, more grounded, to be sure, than that of Gall and Spurzheim, but still at best providing a description, not an explanation. At quite another level, proteomics is likely to provide another, entirely different, cartography. But we don't know how the two, the proteomics and the imaging, relate, or how either and both change over time. There are multiple spatial and temporal scales and dimensions engaged here.

One way now being advocated of moving between levels would be to combine brain imaging with single-cell recording by means of electrodes placed within or on the surface of individual neurons. These procedures cannot be ethically performed on humans except for well-defined medical purposes. However, they could in principle be allowed on primates. This was one of the scientific arguments being advanced for the controversial proposal to build a primate centre in Cambridge, but eventually shelved because it was judged that the costs of protecting it from angry animal-righters would be too great. The claimed justification for the Centre, that it would aid in research on Alzheimer's and Parkinson's diseases, was to my mind always pretty spurious. Whilst I agree that the combination of single cell recording with imaging would help move our understanding of brain mechanisms beyond pure cartography, I share with many the sense of unease about whether this end justifies these particular means – the use of animals so closely kin to humans. I've no doubt that such studies will be pursued elsewhere in countries less sensitive than Britain about the use of animals. It would then perhaps be possible to test the idea that combining imaging and single-cell recording will increase our understanding of global neural processes such as those that enable the brain to solve the binding problem. But if critics like Walter Freeman are right, and what matters is not so much the responses of individual cells but the much wider field effects of current flow across their surfaces, then an essential element in interpretation will still be lacking.

A second approach with powerful advocates is to use the now common techniques of genetic manipulation in mice, systematically knocking out or inserting particular genes and observing the conse- quences on development, anatomy and function. I don't have quite the same ethical problems about working with mice as I do about primates – I'm not going to spend time here explaining why – but I fear that the results of this approach may produce more noise than illumination. Brain plasticity during development means that if an animal survives at all, it will do so as a result of the redundancy and plasticity of the devel- opmental system which ensures that as far as possible the gene deficit or surplus is compensated for by altering the balance of expression of all other relevant genes. The result is that often, and to their surprise, the geneticists knock out a gene whose protein product is known to be vital for some brain or body function, and then report 'no phenotype' – that is, they can see no effect on the animal because compensation has occurred. But the converse is also true. Because many genes also code for proteins that are involved in many different cell processes, knocking the gene out results in a vast range of diffuse consequences (pleiotropy). Once again, dynamic complexity rules.

Many neuroscientists argue that there is a greater priority than acquiring new data. The world-wide effort being poured into the neuro- sciences is producing an indigestible mass of facts at all levels. Furthermore the brain is so complex that to store and interpret these data requires information-processing systems of a hitherto undreamed- of power and capacity. It is therefore necessary to build on the experi- ence of the human genome project and invest in one or more major new neuro-informatic centres, which can receive new data in appro- priate forms and attempt to integrate them. This seems an innocent, if expensive new project, likely to find favour in Europe, where such a centre is under active consideration as I write. But once again a small whisper of scepticism sounds in my ear, and if I listen hard enough I can hear the fatal acronym GIGO – garbage in, garbage out. Unless one knows in advance the questions to be asked of this accumulating mass of data, it may well prove useless.

This is because behind all of these potentially laudable moves forward there lies a vacuum. Empiricism is not enough. Simply, we currently lack a theoretical framework within which such mountains of data can be accommodated. We are, it seems to me, still trapped within the mechanistic reductionist mind-set within which our science has been formed. Imprisoned as we are, we can't find ways to think coherently in multiple levels and dimensions, to incorporate the time line and dynamics of living processes into our understanding of molecules cells and systems.

Bring on the cerebroscope

So let's try a thought experiment, and tackle what philosopher David Chalmers calls 'the hard problem', the division between objective, third person, and subjective, first person, knowledge and experience.[33] Chalmers, Colin McGinn and others worry about what they term *qualia* – that is those aspects of subjective experience such as the sensation of seeing red. How, they ask, following earlier generations of philosophers, can brain stuff, the neural firing or whatever which an outsider can measure 'objectively', generate such a first-person, subjective experience? Experiencing the redness of red seems to belong to an entirely different universe – or at least an entirely different language system – to statements about neural firing.

It may be because I am philosophically tone-deaf, but I have never found this a very troubling question. It is surely clear that, granted enough knowledge of the visual system, we can in principle, and to some extent in practice, identify those neurons which become active when 'red' is perceived. (Indeed in animal experiments such neurons have already been identified.) This pattern of neural activity translates into the seeing of red, and seeing red is simply what we call in mind language the phenomenon that we call in brain language the activity of a particular ensemble of neurons. This doesn't seem harder to understand than is the fact that we call a particular small four-legged furry mammal *cat* in English and *gatto* in Italian; the two terms refer to the same object in different and coherent, but mutually translatable languages. No problem. But can we go further? Let's imagine that we have all the techniques and information-processing power that neuroscientists can dream of, and consider a theoretical machine – let's call it a cerebroscope (a term which, I believe, was invented many years ago by the information scientist – and explicitly Christian anti-determinist – Donald Mackay) – that can report the activities at any one time of all the 100 billion neurons in the brain. Define this activity at any level – molecular, cellular, systems – that seems appropriate. Now consider a variant of the experience that I discussed in Chapter 6, that we call subjectively 'seeing a red bus coming towards me'.

Our cerebroscope will record and integrate the activity of many neurons in the visual cortex, those that are wavelength sensitive and report red, those that are motion sensitive that report directional movement, edge-detecting neurons, neurons tuned to binocularity etc, all of which combine, via some solution to the binding problem, to create an image of an object of a given shape and volume, moving towards us with a speed I can estimate from the rate of change the image subtends on my retinae. Acoustic information is also bound in so we can register the engine noise of the approaching bus. But hang on – how do I know that the noise is of an engine, or the object a bus? There must be bound in with all the sensory information some other neural activity which scans and extracts the recognition memory which defines the object as a bus, and the noise as that of an engine. Perhaps the memory involves inferior temporal cortex, and the naming of the bus will engage Broca's area as we identify it as a 'bus'. But let's go one step further. Is seeing this bus a good thing or a bad thing? If I am on the pavement waiting for it, a good thing; if I am crossing the road and it is coming at me fast, a dangerous thing. There is affect associated with these images. Amygdala and ventromedial frontal cortex are involved. Then I must decide how to act – prepare to enter, or jump out of the way – perhaps right parietal cortex and frontal lobe engagement? The appropriate muscles must be engaged, blood circulation adjusted, and so on and so on.

The cerebroscope will enable an observer to record all this activity over the few seconds during which I am perceiving the bus and acting on my perception, and such an observer is entitled to say that the sum total of this activity represents in brain language, my mental processes of seeing, etcetera, the bus. So, once more, what's the problem?

Consider the process I have just described from the other end. Suppose the cerebroscope stores all this information in its gigaterabyte information processor. Then, at some later time, an experimenter asks the machine to present the data and translate it back into mind language, that is, to deduce from the neural activity the thought and action processes that it represents. Could it interpret all the data and print out a statement saying 'What the person, Steven Rose, associated with this brain, is experiencing is a red bus coming towards him and that he is in danger of being run over by it'?

The answer seems to me almost certainly to be no. The interpretation of the firing pattern of any particular neuron is very much dependent on its history. Plasticity during development may mean that even the wavelength to which any particular neuron is sensitive may vary from individual to individual, so what ends up as one person's 'red' neuron may in another person's brain respond to blue and not to red. Even more sure is that whatever is the pattern of neural firing and connectivity in my inferotemporal cortex that corresponds to *my* recall or recognition memory of a bus will not be the same as the pattern in *yours*, even though the outcome – recognising a bus – will be the same in both cases. This is because your and my experience of buses, and how we each store that experience is inevitably both different and unique to each of us. So for the cerebroscope to be able to interpret a particular pattern of neural activity as representing my experience of seeing the red bus, it needs more than to be able to record the activity of all those neurons at this present moment, over the few seconds of recognition and action. It needs to have been coupled up to my brain and body from conception – or at least from birth – so as to be able to record my entire neural and hormonal life history. Then and only then might it be possible for it to decode the neural information – but only if there were then a one-to-one relationship between the history and present state of my neurons and my mental activity. And this we simply don't know. There may be an indefinite number of histories of neurons from conception to the present time which could be interpreted as meaning the experiencing of a red bus coming towards me – and equally there might be an infinite number of experiences that could be inferred from any particular pattern.

As a Christian and troubled by the free-will/determinism problem, Mackay was interested in a further question. Could the information obtained by his hypothetical cerebroscope enable an observer to predict what the observed person would do next? He considered the cerebroscope as being able to report back to a person the state of her own brain and therefore mind at any time, because she as the person being observed can also receive reports from the machine. But, as he pointed out in an argument against determinism and in favour of some form of 'free-will', this raises a paradox, for the act of reporting back to a person would change the state of their brain in unpredictable ways, and hence the predicted outcome would itself be modified. In a hugely simplified manner this is, of course, just what the biofeedback devices that have been available for some years now to help people reduce stress or learn to meditate can do. Thus, even if we knew everything we needed to know subjectively/mentally about the 'objective' state of our brain at any one time, our actions would not therefore be determined. I don't really think that this provides the solution to the 'free-will paradox', for reasons I've discussed earlier, but it does explore the limits to how any understanding of the brain might help us to understand the mind.

On imagination

Seeing and responding to an oncoming bus might seem pretty straightforward, but St Augustine's challenging questions go much deeper. He asks, for instance, how we can remember colours or feelings in the absence of the stimuli that generate them. What would happen if, instead of studying the brain processes that occur when I am crossing the road in front of the bus, I simply imagine that this is what I am experiencing, or remember a past real experience? Once we understand the language in which memory is encoded within the brain, and how recall occurs, it doesn't seem impossible to account for how we can remember the colour red in its absence – by activating the relevant 'red' neurons. And memory of feelings could similarly evoke activity in those neurons of the amygdala and frontal cortex and their interconnections which are engaged in the original sensation of feeling. There is a growing body of evidence from imaging studies suggesting that this is indeed what happens. Told to think about a cat and then asked about some specific feature – for instance, the shape of its ears – specific regions in the primary visual cortex become active.[36] Teach a person a motor activity

– say a finger-tapping sequence – and regions of the motor cortex are active when the movements are performed. Asked to imagine the activity without performing it, the same regions (though not necessarily the same cells) will become active. Asked to think about a familiar journey – for instance from one's house to the shops – dynamic patterns of electrical activity across the cortex track the mental journey as one turns first left, then right. So mental activity in the absence of non-mental stimuli can result in similar if not identical neural activity as that generated by the material stimuli. Then there are the 'mirror neurons' I discussed in Chapter 6 – neurons that fire both when a monkey is performing a particular task or when it observes another performing the same task.[37] Whether the cerebroscope could distinguish between our imagined or empathised experience and the real thing is an open and very interesting question, not far beyond the limits of current research to discover.

How about Augustine's less material examples, such as 'abstract propositions, the principles of numbers and dimensions . . . false arguments . . . the idea of God?' As Stanislaus Dehaene has shown, there are regions of the brain that become active when people are asked to solve mathematical problems, and there are distinct individual differences in brain activity between those with weak and strong mathematical skills.[38] Others have spoken of identifying a 'God centre' in the brain – presumably a region that is active when a believer thinks about his or her personal deity (try looking up God Centres in the Brain on Google for these claims). Perhaps this shares a location with Augustine's 'false arguments', as there are also claims to be able to distinguish on the basis of brain signals between 'true' and 'false' memories.[39]

A little caution is necessary before we claim that neuroscience can solve Augustine's problems. Bring back the cerebroscope, even one empowered to record a person's neural activity from the first moment of detectable neural activity in the foetus onwards, and focus it on the brain of someone trying to decide whether an argument is false or not. Once again we will expect all sorts of brain regions to light up as some proposition is examined, semantically, syntactically, compared with related propositions extracted from memory, and so forth. The cerebroscope will in due course also register the final decision, yes or no, to the truth or falsity of the proposition – but would it be able to detect the actual *content* of the argument leading to the conclusion? I suggest not; the cerebroscope is at the limits of its powers in identifying the brain regions that *enable* the mental process involved in the argument. It is at this point, I suggest, that neuroscience may be reaching its theoretical limits in its efforts to understand the brain in order to explain the mind.

Of course, our cerebroscope is imaginary, the figment of my own mind/brain, for in its final manifestation, one that acts as a recording angel for every moment of our brain's life, it is being asked to perform an impossible task. To describe and analyse mental events, streams of thought, the 'feelings' which Damasio contrasts with 'emotions', we habitually use the form of mental communication we know best – that of language. We can describe our feelings, our memories, our dreams, the logical sequence by which we decide on the truth or falsity of a proposition, in a symbolic system which evolution has made uniquely human – that of language, spoken or written. We can use this language subtly, powerfully, persuasively, analytically and systematically, encompassing not just the present but past and future too, in ways in which even the best of cerebroscopes will never be able either to interpret or compete.

Some years ago at a symposium on reductionism in the biological sciences I clashed with the distinguished philosopher Thomas Nagel on this matter of levels of explanation. Arguing for the primacy of reductionist explanations, Nagel suggested that whilst higher level accounts, such as mentalistic ones, could *describe* a phenomenon, only the 'lower level' reductionist ones could *explain* it. I disagree. In many cases, lower-level accounts are descriptive whilst the higher-level ones are explanatory. However comprehensive the cerebroscope's record of the neural activity taking place when I experience the sensation of being angry or in love, drafting this sentence or designing an experiment, the account will only be descriptive. It is the words, read or spoken, which are explanatory. For sure, explaining the brain is helping us describe and understand our minds, but it is not going to succeed in eliminating mind-language by relocating it into some limbo dismissable by the cognitive illuminati as mere 'folk psychology'.

However, and it is to these issues that I must finally turn, even our limited capacity to explain brains is already providing powerful tools in the attempts to heal, change, modify and manipulate minds.

CHAPTER 9

Explaining the Brain, Healing the Mind?

The emergence of mental illness

THROUGHOUT RECORDED HISTORY, ALL CULTURES HAVE BEEN ABLE TO recognise mental anguish and despair, those conditions that we now call anxiety and melancholy, disorientation and hallucination, or even 'simplemindedness'. Accounts exist in all the ancient texts. Western traditions, following the Greeks, related human temperament to mixtures of four basic humours, analogous to the four elements of air (dryness), fire (warmth) earth (cold) and water (moisture), related in turn to yellow bile, blood, phlegm and black bile. Mental distress and disorientation was seen as indicating a disproportion in this metaphorical mix. Not dissimilar concepts of disharmony, imbalance in vital forces or juices (Yin/Yang), especially concerning the heart, characterise traditional Chinese thinking.[1] Treatments (apart from forms of exorcism to drive out the evil spirits) consisted of various combinations of herbal extracts intended to restore harmony and balance.

In pre-industrial agricultural societies, eccentricities of behaviour and simpleness were probably more tolerated and accommodated within the community, just as senescence was recognised as involving a loss of mental powers. Hamlet may have feigned madness, but the aged Lear cries out to heaven 'let me not be mad'. Many forms of behaviour today regarded as 'abnormal' or as an aspect of a diagnosable disease or syndrome were almost certainly regarded in the past as merely part of the rich tapestry of human variation. Classifying such forms of distress as diseases – still less brain disorders – is much more recent. Industrialisation and urbanisation are less tolerant. Many historians, with widely different perspectives, have charted the phenomenal rise of the asylums for the insane from the eighteenth to the mid-twentieth century.

Recognisably modern attempts to distinguish between different forms of mental disorder, to classify and label symptoms, really begin in the late nineteenth century. The historical debt owed to the tradition of splitting mind and body is transparent, and even today a distinction is often made between 'neurological' disorders which involve overt brain lesions – including in this context Parkinson's and Alzheimer's diseases – and 'psychological' problems where there are no such obvious signs of brain damage. The distinction is not merely conceptual but enshrined in practice, as different medical specialities (neurologists versus psychiatrists) are involved, and there are important turf wars between the disciplines.

Based on post-mortem examination of the brain, some diseases could be transferred from the psychological to neurological. Thus Korsakoff's syndrome, the shrinkage of the brain associated with acute alcoholism, was recognised as proximally caused by the resulting deficiency of thiamine, one of the B vitamins. An entire subset of mental problems associated with diseases like goitre and beri-beri began to be attributed to vitamin deficiencies. 'General paralysis of the insane' could often be the consequence of progressive syphilis. Dementias, previously distinguished only as 'senile', 'paralytic' and so forth, were reclassified by Théodule Ribot in his classic text *Les Maladies de la Mémoire* in 1881 according to a temporal sequence which begins with loss of memory for recent events and eventually leads even to the loss of normal bodily control. In 1906 Alois Alzheimer first described a case of progressive dementia in a woman whose brain, on post-mortem, showed the characteristic degeneration and plaques now seen as diagnostic of Alzheimer's disease.

By the end of the nineteenth century, attempts had also begun to categorise more precisely the forms of mental distress that could not be associated with overt brain damage – a task usually associated with the name of Emil Kraepelin. The main distinction he drew was between apparently transient disorders and distress (such as the affective disorders, depression, anxiety and bipolar disorder) and those that seemed permanent. These persistent disorders of thought, of hallucinations and paranoia, were named schizophrenia by Eugen Bleuler in 1911. Kraepelin's and Bleuler's classifications have dominated psychiatry ever since, despite doubts as to the validity of the distinctions. Schizophrenia, often manifesting itself in people in their late teens and early twenties, is said to be characterised by 'thought insertions', delusions and auditory hallucinations, and the sensation that one's thoughts and actions are being controlled by alien forces outside one's control. It is said to affect between 0.5 and 1 per cent of the population, but is twice as commonly diagnosed in working-class than in middle-class people, and notably in Britain is excessively diagnosed in people of Caribbean origin and in children of so-called 'mixed-race' relationships.[2] Its existence as a diagnosable entity has been denied by some, and by others, especially during the height of the 'anti-psychiatry movement' of the 1960s, it was regarded as a sane response to impossible circumstances,[3] whilst its diagnosis has been criticised as representing a means of social and political control.[4]

The clearest cut demonstration of the problematic nature of such psychiatric diagnoses comes from a sensational experiment carried out by David Rosenhan in 1973 in California, published with the provocative title 'On being sane in insane places'.[5] He and a group of eight volunteers each checked themselves in at different hospitals, claiming that they heard voices repeating the word 'thud'. Each was admitted into psychiatric care, after which their instructions were to behave perfectly normally and explain that the voices had ceased. Their 'normal' behaviour was interpreted as 'abnormal' (taking notes was described as 'compulsive writing behaviour'), but it took them a remarkably long time, despite their now 'sanity', to be discharged as 'in remission'. The enraged psychiatric profession claimed Rosenhan had fooled them, so, in a follow-up experiment, he announced that another group of 'pseudo-patients' would be presenting themselves at one of the hospitals. In the subsequent month the hospital psychiatrists declared that they had identified forty-one such pseudo-patients and admissions dropped substantially. In fact, there were no pseudo-patients. Frightened at having been fooled, the psychiatrists had fallen even further into Rosenhan's trap.

Well, that was then, and here we are thirty years later, so it couldn't happen again, could it? In the 1970s the US was diagnosing schizophrenia at a much higher rate than any other country;[6] the only nation

with as wide diagnostic criteria was the Soviet Union, which was casti-
gated for using the label as a way of dealing with political dissidents, by
incarcerating them in prison hospitals.[7] The US zeal has somewhat
abated, and its diagnosis rate is now no different from Europe. So in
2003 Lauren Slater repeated Rosenhan's experiment, this time as a solo
act, but checking herself in at a number of different hospitals in
California. Unlike Rosenhan's pseudo-patients, she wasn't admitted to
hospital, but was variously diagnosed as being psychotic or depressed
or suffering from post-traumatic stress disorder. In each case she was
prescribed powerful drugs, including the anti-psychotic Risperidol.[8]
Such experiences lend force to the arguments of those psychiatrists, like
Richard Bentall, who oppose the practice of shoehorning mentally
distressed people into tidy diagnostic categories.[9] Indeed, anyone with
direct personal experience of living with distress is likely to feel less than
certain that such neat diagnostic criteria properly match their own condition or that of a person they are caring for.

Yet, despite this, in the century following Kraepelin and Bleuler, the
net of mental distress and disorder has been cast ever wider as the
psychiatric community seeks to classify as aberrant, or diseased, more
and more aspects of the human condition. Like schizophrenia, bipolar
affective disorder (manic depression) is now said to affect 0.5–1 per cent
of the world's population, and 'personality disorders' another 5 per cent.
A range of ever more subtle conditions have found their way into the
successive editions of the psychiatrists' bible, the *US Diagnostic and
Statistical Manual* (DSM), ranging from Dissociative Identity Disorder
through Anti-social Personality Disorder to Conduct and Oppositional
Defiance Disorders; I will return to these in later chapters. (On the other
hand, homosexuality, classified in earlier versions of the DSM as a
psychiatric disease, has now been removed.)

By far the broadest disease categories, now regarded by the World
Health Organisation as a global epidemic, are those of depression and
anxiety (affective disorders). Forty-four million Americans are said to be
suffering from depression, with women three times as likely as men to
be diagnosed with it. Estimates of its prevalence range widely, with up
to 20 per cent of the world's population said to be clinically depressed,
and the figures are steadily increasing.

What are the origins of this epidemic? How can it be that the three
billion years of evolution that have generated the human brain have

produced an instrument seemingly so superb, yet also so readily disturbed, so easily made dysfunctional? Is this vast psychiatric caseload the inevitable downside of the creativity and capacity for affect and cognition that so conspicuously characterises our species? For instance, there have been suggestions that schizophrenia is the genetic flip-side of the mutations that generated the human capacity for language – even though the genetics of schizophrenia are, to say the least, ambiguous. Or is it, as some conspiracy theorists (such as scientologists) suspect, a medicalising myth through which people are kept in thrall by a sinister psychiatric establishment? Have our minds and brains simply become out of kilter with, unable to cope with, the complexity of modernity, the pace of technological change and globalisation? These questions will occupy me in the next two chapters, but before we get there it is important to chart the ways in which the descriptions of mental distress and disorder – at the interface as they are between neuroscience, psychiatry and medicine – segue into the attempts to treat and cure the conditions, and to control and modify brain and behaviour.

From description to treatment

By the beginning of the last century, the search for therapies for mental disorders, and 'scientific' methods of moulding minds and brains, were forming the buds that have opened into today's rich flowering. Mere reading of what was given by nature's chance and necessity had become insufficient; the claim was that by the appropriate application of psychological methods, character and capacity could be endlessly manipulated. As the founder of behaviourist psychology in the US, John B Watson put it in an oft-quoted and as oft-derided passage:

> Give me a dozen healthy infants, well-formed, and my own specified world to bring them up in and I'll guarantee to take any one at random and train him to become any type of specialist I might select – doctor, lawyer, artist, merchant-chief and yes, even beggar-man and thief, regardless of his talents, penchants, abilities, vocations, and race of his ancestors.[10]

Steven Pinker takes this as the archetype of 'blank slate' thinking.[11] However, I can't help reflecting that Watson's claim is after all not so different from the way the class and educational system has tended to operate. The British aristocracy, with the eldest son as the heir to land and title, traditionally put the next into the military, and the third into the church. After all, Watson doesn't claim that the adults he reared would make *good* doctors, or *successful* merchant-chiefs, merely that he could direct their interests and career choices. In any event, with his clear gift of persuasion it is not surprising that he soon abandoned academia for the presumably more lucrative world of advertising. Watson's more famous and academically influential successor was BF Skinner, whose mechanistic view of human behaviour led him to argue that humanity's animal nature rendered obsolete such concepts as human freedom and dignity. By contrast, he proposed a future society in which careful conditioning procedures in the hands of an ethical elite would usher in Utopia.[12] Skinnerian methods of 'shaping' behaviour by what he called neutrally 'contingencies of re-inforcement' – in effect, rewards and punishments – became the basis for behavioural therapy in the hands of psychiatrists, and, in the hands of prison authorities and some schools, 'token economies' for controlling the behaviour of inmates. In such regimes, subjects earn or lose points for carefully defined good or bad behaviour, and can trade in the points for privileges, thus presumably directing – shaping, in Skinnerian language – their future conduct.

Behaviourism was based on a peculiarly naïve view of how not just people but animals in general behave, a view of organisms as entirely mechanistic, born empty of predispositions, lacking, to quote Watson, talents or penchants. All were assumed to be acquired by experience during development – experience of which types of action produced rewards from the environment and which resulted in punishments. The chosen 'environment' for experimentally-minded Skinnerians was a box devoid of anything except levers or coloured symbols that could be pressed if you were a rat or pecked if you were a pigeon to obtain food or drink. The trouble for such research is that animals do have predispositions, and don't always interpret the intentions of the experimenter in the 'right' way: for instance manifesting so-called superstitious behaviour in which they behave as if they believe they will be rewarded if they perform some action quite unconnected with that intended by the experimenter. As I've said, by the 1960s this 'mis-behaviour' of animals in more naturalistic environments helped sound the death-knell of behaviourism in the laboratory, if not the clinic, the school or the prison.

With the advent of newer technologies, however, came the prospect of more direct physical intervention into neural processes, always of course in the interests of the patient. A forerunner was the Portuguese neurologist Egas Moniz, who in the 1930s developed a procedure for the treatment of schizophrenia and related conditions which involved removing the entire frontal lobe (lobotomy) – later refined into doing no more than severing the tracts that connected it with the rest of the cortex (leucotomy). Moniz's patients were said to be calmer and more tractable after such treatment, and he was awarded the Nobel Prize for it in 1949. The method was taken up enthusiastically in Europe and the US. By 1949 some 1200 patients a year were being lobotomised in Britain, whilst in the US the flamboyant Walter Freeman* demonstrated his skill at leucotomising unanaesthetised patients, using an ice-pick driven through the orbit of the eye, in front of – presumably – admiring student audiences.[13]

These 'heroic' procedures (heroism is often ascribed to doctors and surgeons where it might better be restricted to their patients) declined towards the end of the 1950s, when newer generations of psychoactive chemicals began to be developed. However, psychosurgery, as it came to be called, lived on in other forms, especially in the US in the treatment both of patients in mental hospitals and of prison inmates. Thus there was considerable interest in the observation that removing the amygdala in rodents rendered them less 'aggressive'. Aggressivity is measured in rats by trading on their propensity to kill mice. A rat is placed in what is essentially a large empty fish tank, and in due course a mouse is dropped in alongside. The time taken for the rat to kill the mouse is taken as a measure of its 'aggressivity'. The quicker the kill, the greater the aggression. In fact, it's a pretty ropey measure. It ignores entirely the past history of the two animals, the order in which they are placed in the tank, and the general environmental context. For instance, if the

* Not to be confused with his son, the neurophysiologist Walter J Freeman, whose work on chaotic dynamics in the explanation of mental activity I discussed in Chapter 6.

rat and the mouse are familiar with one another, perhaps having been reared together, or if the environment is more complex, the rat is far less likely to strike.

Be this at it may, it was not too long before these rattish observations became extrapolated to humans. Faced with the escalating violence in US cities during the 1960s, two US psychosurgeons, Vernon Mark and Frank Ervin, in research funded by US law enforcement agencies, argued[14] that the riots may have been precipitated by individuals with damaged or overactive amygdalae, and that a potential treatment would be to amygdalectomise selected 'ghetto ringleaders'. Some 5–10 per cent of US citizens, they claimed, might be candidates for such psychosurgery. (This magic percentage range will crop up several times in what follows.)

An indication of the candidates proposed for such operations is provided by an exchange of correspondence between the Director of Corrections, Human Relations Agency (*sic*), Sacramento, and the Director of Hospitals and Clinics, University of California Medical Center, in 1971. The Human Relations expert asks for a clinical investigation of selected prison inmates 'who have shown aggressive, destructive behavior, possibly as a result of severe neurological disease' to conduct 'surgical and diagnostic procedures . . . to locate centers in the brain which may have been previously damaged and which could serve as the focus for episodes of violent behavior' for subsequent surgical removal.

An accompanying letter describes a possible candidate for such treatment, whose infractions whilst in prison include problems of 'respect towards officials', 'refusal to work' and 'militancy'. He had to be transferred between prisons because of his 'sophistication . . . he had to be warned several times . . . to cease his practicing and teaching Karate and judo. He was transferred for increasing militancy, leadership ability and outspoken hatred for the white society . . . he was identified as one of several leaders of the work strike of April 1971 . . . Also evident at approximately the same time was an avalanche of revolutionary reading material.' To which request the Director of Hospitals and Clinics replies, offering to provide the treatment 'on a regular cost basis. At the present time this would amount to approximately $1000 per patient for a seven-day stay . . .'[15] Cheap at the price.

There is a degree of crudity about such procedures; neurotechnology was after all only in its infancy at the time, and already less invasive measures were becoming available, based on the technique of permanently implanting radio-controlled stimulating electrodes into specific brain regions. These were elegantly demonstrated by the neurophysiologist Jose Delgado in the 1960s. In a bravura and still unexcelled display of showmanship, he implanted electrodes into the brain of a 'brave' (that is, aggressive) bull, ostensibly into the region controlling 'aggressivity' (perhaps the amygdala, although his account does not specify), donned a toreador's costume and cape and, carrying a radio transmitter, entered the bullring. A filmed sequence shows the charging bull being brought to a halt when Delgado switches on the current. The power of the demonstration was somewhat weakened by suggestions that the electrodes were actually in the motor cortex and thus inhibiting movement rather than aggressive intention, but this did not prevent Delgado arguing for the prospective use of this technology in humans. As he put it, arguing for what he called a 'psychocivilised society': 'The procedure's complexity acts as a safeguard against the possible use of electrical brain stimulation by *untrained or unethical persons* [my italics].'[16]

This elision of expertise with ethics requires no further comment.

Pharmacology to the rescue?

The use of extracts or ferments of fungi or plants, from magic mushrooms and mescal to cannabis and alcohol, to affect mood and perception is certainly older than recorded history. Many modern psychochemicals are derived from plants, or synthetic substances based on naturally occurring molecules often based on variants of traditional herbal remedies. A classic example of such plant-derived products is aspirin, a derivative of salicylic acid, present in the bark of willow trees (*Salix*, in the Latinate classification), discovered, extracted and marketed by the German company Bayer at the turn of the last century. Reserpine, a major tranquilliser, was purified from the Himalayan shrub Rauwolfia, a plant that figures prominently in the Indian (Aryuvedic) pharmacopeia as a treatment for agitation and anxiety. Extracts of the common European plant St John's wort are now a popular and seemingly effective 'natural' alternative to prescription antidepressants.

In general, modern Western medicine has been ambivalent towards such traditional remedies, on the one hand dismissing them as quackery and on the other studying them as a potential source of patentable products. Scepticism is muted by the simultaneous prospect that ignoring such folk remedies from many disparate societies, may mean missing out on potentially novel drugs, the successors to reserpine and its fellows. So the pharmaceutical companies employ 'drug prospectors' to trawl the world from the Andes to the Himalayas for plants with claimed psychotropic effects, with the aim of purifying, patenting and controlling the supply of the active agents they may contain. Hence the ambivalence towards claims for such substances as ginseng, ginko, echinacea and the many others that now pack the shelves of health food stores and the alternative medicine market.

Many are none the less dismissed on the grounds that they fail to demonstrate their effectiveness when tested under standard scientific procedures such as double blind random control trials. This is the procedure in which groups of patients are given either the drug or an inert placebo (or, sometimes, a drug with already proven effect), without either the clinician or the patient knowing which is which. The effects of the two can then be compared. This has become the 'gold standard' in drug testing and in so-called 'evidence-based medicine'. It sounds sensible and so indeed in many but not every context it is. Whilst all new drugs brought on to the market must go through such trials for efficacy and safety before being licensed, the same is not necessarily true for the traditional phytopathic (plant) and naturopathic remedies available without prescription. Thus they fall outside the control of the medical and pharmaceutical authorities, and, being prepared as complex extracts rather than pure chemicals, may vary substantially in strength and indeed chemical composition, thus adding to the uncertainty of the clinical trials. These are good reasons for medical suspicion.

Such trials are not without other problems, even with chemically synthesised substances. A fair proportion of patients, especially those suffering from depression, show significant recovery with the placebo, whilst some on the drug regime show no improvement or even adverse reactions. 'Adverse reactions' used to be, and sometimes still are, called 'side effects'. The term implies that the drug is a sort of magic bullet with only a single target. Assuming a unique target is analogous to the military thinking that offers us such euphemisms as 'smart bombs' and 'collateral damage'. Putting any new chemical into the body is the real equivalent of a spanner in the works, and is likely to have multiple effects – both wanted and unwanted – on many enzyme and cell systems.

Even in the absence of such reactions, there are problems in interpreting drug trials. One is the assumption that a diagnosis like 'depression' refers to a specific entity rather than being a label given to a spectrum of different states of mind, each with associated but not necessarily similar changes in body and brain chemistry. A second and increasingly well-recognised problem is that people are not biochemically identical. Genetic and developmental variation amongst individuals means that not everyone will respond to the same drug in the same way. The research director for one of the biggest of the pharmaceutical companies, Allan Roses of GlaxoSmithKline, caused a mild furore in 2003 when he pointed out that at best drugs 'worked' in only about half the patients who were prescribed them, whilst adverse reactions occurred in many cases. The reasons why he made such a seemingly damaging confession, no surprise to those in the trade, will become clearer in the course of the next chapter.

From barbiturates to Valium

The birth of the modern psychotropic industry – today a $49 billion annual market – can probably be dated to 1912, when Bayer, perhaps emboldened by its success with aspirin, marketed the drug phenobarbital (Luminal) which acts as a sedative, prescribed to calm – indeed to put to sleep – agitated patients. Phenobarbital and its relatives (Amytal, Nembutal, Seconal, Pentothal) are effective anaesthetics and are indeed sometimes still used as such. They have multiple effects, including depressing cerebral energy metabolism and electrical activity. They soon came to be generally prescribed for sleeping problems – by 1970 there were some twelve million prescriptions a year being written for them in Britain. However, the barbiturates proved problematic. Putting an alien chemical into a highly homeodynamic biochemical system results in the system responding by producing compensatory changes in enzyme levels and metabolic processes, minimising the effect of the chemical spanner in the works. This adjustment is called tolerance and means that higher and higher doses of the drug need to be used to achieve the same physiological/mental effect. The result of modifying biochemistry to respond to the drug has a further effect though. If the drug is abruptly discontinued, its absence is noticed and extreme physiological distress can result. This is the biochemical cause of addiction. Taking barbiturates can result in both tolerance and addiction. There is a further and major problem. The margin between a sedative dose and a lethal dose of a barbiturate is not so great as to prevent them becoming a favoured means of assisted death. Whilst they are still prescribed, they are now recommended only in treating severe insomnia.

The limited utility of the barbiturates led, from the 1950s on, to an increasing search amongst drug companies for more specific and effective tranquillising agents. The first major success came with Rhône-Poulenc's chlorpromazine (Largactil), which reduces hyperactivity in manic patients, above all in those diagnosed with schizophrenia. Like many such drugs, its effects on brain and behaviour were discovered almost serendipitously – it had originally been synthesised in the hope that it might act as an anti-histamine. Chlorpromazine was introduced into Britain in 1952, and is claimed to have revolutionised psychiatry, opening the locked back wards of the mental hospitals and symbolising the birth of the modern psychotropic era. It was initially seen as an almost universal elixir, ranging from the relief of depression through to:

> The allaying of the restlessness of senile dementia, the agitation of involutional melancholia, the excitement of hypomania . . . the impulsiveness and destructive behaviour of schizophrenics. The patients develop a kind of detachment from their delusions and hallucinations . . . especially striking is the effect on many cases with paranoid delusions, whose persecuted distress may be dissolved . . .[17]

Within ten years of its introduction, it is estimated that chlorpromazine had been given to 50 million people world-wide. Quite how a single chemical could have such widespread and dramatic effects – a true 'magic bullet' – was unknown. However, before long reports began to accumulate of a troubling consequence of its long-term use – patients began to develop increasing bodily rigidity, coupled with twitching and abnormal movements of hands and mouth, including involuntary thrusting out of the tongue. The condition – tardive dyskinesia – did not remit when the drug was removed. Prolonged use of chlorpromazine causes irreversible damage to neural systems using the transmitter dopamine; in effect, its use results in something akin to Parkinson's disease. To say the least, this finding diminished the enthusiasm for its use, and it is no longer a drug of first call, although as the prescriber's bible, the *British National Formulary*, dryly puts it, it is 'still widely used despite the wide range of adverse effects associated with it . . . useful for treating violent patients without causing stupor'.

Elucidating chlorpromazine's biochemical mode of action and the role of the dopamine receptors helped bolster the argument that schizophrenia and bipolar (manic-depressive) disorder, along with depression and anxiety, all involve some disturbance in neural transmission at synapses. Despite increasing doubts as to the simplicity of 'the dopaminergic hypothesis' of schizophrenia, new drugs, such as fluphenazine and haloperidol, interacting more specifically with particular dopamine receptors, soon followed in chlorpromazine's wake, managing thereby to avoid the adverse effects – the tardive dyskinesia – induced by chlorpromazine.

The discovery that chlorpromazine's effects seemed to be mediated via the dopaminergic system helped focus the attention of the pharmaceutical companies on the development of drugs that could interact with other neurotransmitter systems. This could involve over- or underproduction or degradation of the transmitter or affect the efficacy of one of its many receptor molecules. Much drug research therefore concentrated on the chemical synthesis of substances whose molecular structure could be predicted to make them likely to interfere with one or other aspect of neurotransmitter function. Virtually all of the current generation of psychotropic drugs are assumed to exert their effects via either stimulating or damping down neurotransmission. Apart from dopamine, the key targets have been the transmitters GABA and serotonin (5-hydroxy-tryptamine).

GABA, as I mentioned in Chapter 6, is the major inhibitory neurotransmitter in the brain, serving therefore to prevent rather than enhance transmission across the synapses. The initial discovery of drugs which interacted with GABA was, however, accidental, and their use in treatment preceded the uncovering of just how they worked at the cellular level. The finding emerged by way of rival drug companies trying to find compounds that could compete with chlorpromazine in their sedating effects. This led to the identification of a class of 'anxiolytic' chemicals – substances that reduced agitation without the heavy sedating effects of either chlorpromazine or the barbiturates, and with less possibility of lethal overdosing. These are the benzodiazepines, of which the best known is marketed as Valium. Introduced in the 1960s, Valium became the next in the long line of wonder drugs, magic pills without which it would become impossible to get through the day – 'Mother's little helpers' – although it wasn't until the mid 1970s that it was discovered that they worked by augmenting the effect of GABA. This made it almost too easy to conclude that the drugs achieved their behavioural effects by slowing down an 'overactive' brain, and that the 'cause' of the anxiety was a deficiency in the amount of either GABA or its receptors at key neural sites. Handed out as readily as smarties in the 1970s, Valium and its relatives too turned out to have hooks, and to produce dependency, if not addiction. Once again, the *Formulary* exudes caution:

> Although these drugs are often prescribed to almost anyone with stress-related symptoms, unhappiness, or minor physical disease, their use in many situations is unjustified. In particular, they are not appropriate for treating depression . . . In bereavement, psychological adjustment may be inhibited . . . in children anxiolytic treatment should only be used to relieve acute anxiety (and related insomnia) caused by fear (eg before surgery) . . . Treatment should be limited to the lowest possible dose for the shortest possible time.[18]

Despite these cautionary words, in December 2003 it was reported that some 50,000 children and adolescents in the UK were being treated regularly with a variety of antidepressant and anxiolytic drugs, often unlicensed for use with young people.[19]

From the discovery of the mode of action and effectiveness of such drugs to the assumption that deficits in the neurotransmitter systems they interact with are the *causes* of the psychiatric conditions that they are prescribed for is but a small and seemingly logical step – one that the psychopharmacologist Giorgio Bignami has called *ex juvantibus* logic.[20] After all, the dopaminergic deficit in the substantia nigra in Parkinson's can clearly be related to the muscular tremors that characterise the disease, simply by a consideration of the pathways in the brain that rely on dopamine as a transmitter. However, the Parkinson example can easily lead one astray. If, for instance, one has a toothache, and taking aspirin reduces the pain, one should not jump to the conclusion that the cause of the toothache is too little aspirin in the brain. Aspirin may block the sensation of pain, and chlorpromazine or benzodiazepine may dampen agitation, without revealing anything about causation. But this causal chain is precisely the *ex juvantibus* assumption, one that has produced some useful drugs, proved extremely profitable for the pharmaceutical companies in the past decades, but has also very much distorted the search for explanations of either the proximal or the distal causes of the conditions – from depression to schizophrenia – that the drugs are intended to treat. Ronald Laing, famous in the 1960s for his part in the anti-psychiatry movement, gave a – perhaps apocryphal – example of this masking effect in his description of a Scottish patient who had been hospitalised hearing voices – auditory hallucinations. Following drug treatment, he was seen roaming the ward, complaining bitterly to his voices, 'Speak up y' buggers, I cannae hear ye.'

Serotonin, causes and correlations

The other major focus of pharmaceutical attention, and the one that by the 1990s and the beginning of the present century had become the latest in the long line of targets for potential magic bullets, has been the serotonin system. Serotonin is a monoamine, an excitatory neurotransmitter synthesised in the raphe nucleus, a region deep in the brain, in the midline of the brain stem, and with axons that project to many other regions, notably the cortex. The raphe is believed to be associated with the perception of pain. Serotonin pathways are involved in many diverse regulatory mechanisms, and the transmitter's effects are mainly achieved in a slightly roundabout way, by modulating the outputs of other excitatory or inhibitory neurotransmitters.

The rise to prominence of drugs affecting the serotonergic system began in the 1950s, when the Swiss firm Geigy (now part of the merged company Novartis) was hunting for a competitor to chlorpromazine. According to the psychiatrist Samuel Barondes, the compound they chose was imipramine. Although ineffective or worse in treating schizophrenia, it turned out to have a strong antidepressant action. Patients '. . . become more cheerful, and are once again able to laugh . . . Suicidal tendencies also diminish . . . sleep occurs again . . . and the sleep is felt to be normal and refreshing.'[21] Imipramine (marketed as Tofranil) is one of a class of molecules built around carbon ring structures and therefore known as tricyclics, and its success heralded the arrival of a family of tricyclic rivals and alternatives.

I've already mentioned a number of ways in which a drug can interfere with the mechanism of action of a neurotransmitter. It can increase the transmitter's synthesis, or inhibit its breakdown. This latter is the way the cholinergic drugs used in Alzheimer's disease work (as do some of the nerve gases used as chemical weapons). One class of antidepressants, the monoamine oxidase inhibitors (MAOIs), which prevent the breakdown of serotonin, work like this. Or, the drug can interact with one or more of the specific receptor sites for the transmitter, like haloperidol for dopamine. However, there is also another, more subtle route. When a neurotransmitter is released into the synaptic cleft, some of it binds to the post-synaptic receptor, but any surplus transmitter will be taken back into the presynaptic cell for re-use or destruction. This is a process known as reuptake. Inhibiting reuptake will increase the amount of transmitter available to interact with the receptor. Imipramine, which binds to serotonin receptors, was the first of the serotonin reuptake inhibitors (SRIs).

For some years imipramine and its close relatives, such as amitriptylene, became the drugs of choice for depression. Furthermore, unlike the drugs discussed earlier, whose direct biochemical effects cannot be studied in humans, a peculiar feature of serotonin receptors makes them accessible. For embryological reasons that don't really matter here, one of the several types of cell present in blood, the platelets, carry the systems required for serotonin reuptake on their membranes, just as do serotonergic nerve cells. Platelets can easily be extracted from a few millilitres of blood, and the activity of their serotonin reuptake mechanisms measured by incubating them with radioactively labelled imipramine and determining how much radioactivity sticks (binds) to the membranes. The level of imipramine binding in platelets thus becomes a surrogate measure for the efficacy of serotonin reuptake in the brain. It is often found to be lower than normal in patients diagnosed with depression, but recovers to a more usual level with therapy. Other potential antidepressants can be compared with imipramine in terms of how strongly they interact with the platelets. The World Health Organisation has adopted the imipramine binding assay as a way of testing the efficacy of new potential antidepressants.

Some years ago, I used this measure myself in an experiment to test if psychotherapy without drugs would also affect imipramine binding – and indeed it turned out to be the case. We worked with a group of therapists and their clients, who entered therapy complaining of depression. A rating measure for depression (called, for its inventor, the Hamilton scale) confirmed that they would have been regarded as clinically depressed. They started therapy with low imipramine binding levels, but within months of starting the therapy both their depression and their imipramine binding levels had recovered to normal levels. Of course, they might also have recovered spontaneously – it is notoriously difficult to devise 'controls' for the psychotherapeutic experience, and many people suffering from depression show improvement over time even when given placebos. However, it is a nice example of the way in which talking can affect the body's biochemistry, just as changes in biochemistry can affect cognitive and affective processes.

There was an interesting sting in the tail of the experiment, as amongst the control groups we worked with was a group of nurses, whose work patterns made them very stressed. Indeed, their imipramine binding levels were only around half the normal range, but when we tested them on the depression rating scale, they scored as normal. Their biochemical measures were out of kilter with how they said they felt.[22]

To believe that the biochemistry is in some way more real or reliable than a person's stated feelings would be a classical example of *ex juvantibus* logic in practice – you may not feel ill, but our measures say you are! There is not and cannot be any straightforward one-for-one relationship between the complexities of our mental experiences and the simplicity of a single biochemical measure. None the less, the last decades have seen a series of such simplistic attempts to correlate the level of a specific biological marker with a psychiatric diagnosis. The fashion for such correlations has followed a rather predictable pattern. Neuroscientists report the discovery of some molecule important in brain metabolism or neurotransmission, and within a short time there is a report that its levels are abnormal in schizophrenic or depressed individuals. Almost every known neurotransmitter and neuromodulator, from glutamate and GABA through to dopamine and serotonin has at one time or another been proposed as 'the cause' of schizophrenia, only to disappear again as some new fashion sweeps the trade.

Of course, one must expect to find that a person's biochemistry alters with their mental state, as shown for instance in our experiment with imipramine binding and depression. However, although many drugs, such as the monoamine oxidase inhibitors, have an immediate effect on some biochemical measures, it takes some days or weeks before there is a noticeable effect on how well a person feels. The cellular and behavioural manifestations are clearly out of synch, for reasons that no one clearly understands. Presumably the relevant correlate to the sense of depression is not the biochemical process most directly affected by the drug, but some downstream metabolic consequence of the presence of the alien chemical.

In any event, a correlation is not a cause. Correlations do not enable one to determine the so-called 'arrow of causation', though they may point to processes of interest. Technically, they are at best 'state markers', indicative of a relationship between biochemistry and behaviour at a given moment in time, but they are readily misinterpreted. Imipramine binding may be low because the person is depressed, rather than the person being depressed because the imipramine binding is low. Many years ago, when I was a graduate student, there was a well-verified claim that certain peculiar metabolites could be found in the urine of hospitalised schizophrenic patients that were not present in control groups. It turned out, however, that these substances were derived from tea, and the patients were drinking much more hospital tea than the control groups.

At least claims for genetic predispositions differ from such ambiguous state markers as, if they are proven, the gene or gene product would be a 'trait marker' – that is, one present prior to and independently of whether, at any particular moment, a person was manifesting depression, anxiety or auditory hallucinations. Hence – especially since the vast increase in genetic research associated with the Human Genome Project – the attention now being paid to the identification of genes associated with 'predispositions' to mental disease. Reports of the identification of such genes have proliferated in the research literature over the past two decades: genes 'for' schizophrenia, manic-depression and anxiety amongst others. Such claims are often loudly trumpeted in the press, only to be quietly retracted a few months later when they cannot be replicated. Despite the fact that schizophrenia and some forms of depression 'run in families', and therefore are likely to have some genetic risk factors associated with them, it remains the case that at the time of writing, no single or small combination of such predisposing genes has yet been reliably identified.

Enter the SSRIs

Imipramine is a serotonin reuptake inhibitor, but it also affects other reuptake systems, notably that for the transmitter noradrenaline. The next step was to develop drugs that were more fussy in their choice of system – the selective serotonin reuptake inhibitors, or SSRIs. The first of these, developed by the drug company Eli Lilly in 1972, was fluoxetine. Marketed under the trade name Prozac, it was first to become famous, and then notorious. Other similar SSRIs produced by rival companies, such as paroxetine (marketed by the company now called, after a series of mergers, GlaxoSmithKline, as Seroxat in the UK and Paxil in the US), soon followed.

In practice, these drugs turned out to be not very different from imipramine in efficacy, but, beginning in the US and soon spreading to Europe and Australia, they soon swept the antidepressant market. By the early 1990s their use had spread far outside the more obviously clinically depressed population and were being widely used in just those contexts against which the *Formulary* advice quoted above applied. In 2002, the world market for antidepressants was worth about $17 billion. Of the total sales for the ten leading therapeutic classes, antidepressants ranked third overall, with a 4 per cent share of the *total* global market for prescription pharmaceuticals.[23] Prozac, according to its advocates, most prominently Peter Kramer in his widely read book, *Listening to Prozac*,[24] made one 'better than well'. The patients he describes become 'more vitally alive, less pessimistic', with better memory and concentration, 'more poised, more thoughtful, less distracted'. Indeed it was not just patients; everyone, it seemed, could benefit from a regular diet of Prozac. Brain serotonin levels had become a metaphor for all aspects of mental well-being, and it had become the magic molecule of the decade.

The euphoria was not to last. Not everyone – indeed only about one in three – patients prescribed Prozac reported improvements, and for many the adverse reactions were severe: sweating, headaches, cramps, weight changes, bleeding, nausea. Furthermore, although not formally an addictive drug, withdrawal from Prozac may produce dizziness, headache and anxiety. Then, even worse, reports began to accumulate of people taking the SSRIs showing violent, murderous or suicidal behaviour. A landmark case was that of Joseph Wesbecker, who, in 1989, soon after starting a prescribed course of Prozac, shot and killed eight of his co-workers at a print factory in Louisville, Kentucky, before turning his gun on himself and committing suicide. By 1994, survivors and relatives sued Eli Lilly for compensation on the grounds that as makers of the drug they had failed to warn of these potential consequences of its use.[25] Although Lilly finally won the case on a split-jury verdict, the trial signalled the beginning of what Kramer calls a backlash against Prozac.

In the US, the veteran campaigner against the use of drugs in psychiatric practice, Peter Breggin, who gave evidence in the Wesbecker case, responded to Kramer's book with one entitled *Talking Back to Prozac*.[26] In it, Breggin marshalled the evidence for these dramatically adverse effects of the drug. Your drug, he insisted, may be your problem, and indeed many of the diagnosed psychiatric conditions may be iatrogenic in origin – that is, caused by the very agents given to treat them by the proponents of biological psychiatry. If indeed the SSRIs could actually increase suicidal feelings or violence, make a person feel worse rather than better, and yet the medical response was to increase the patient's drug load rather than reducing it, this is indeed a serious charge. Also, it is worth emphasising an often ignored fact: iatrogenic mistakes rank high amongst causes of death in advanced industrial societies like the US. For Breggin, as for those psychotherapists uneasy or opposed to drug use, what is needed for anguished patients is not more drugs but more understanding of the person's problems, emotions and feelings.

As evidence began to accumulate of Prozac-associated suicides, compiled by the psychiatrist David Healy,[*27] a group of patients in the US began a class action against GlaxoSmithKline, the makers of Seroxat, which is currently unresolved. Despite these problems the drug regulatory authorities have been slow to respond. The current editions of the *Formulary* add a warning that a consequence of taking SSRIs might be 'violent behaviour', whilst the scandal of prescribing the drug to children, mentioned above, has led to tighter control over their use. The gilt is definitely off the SSRI gingerbread.

---

* David Healy himself became the focus of a major international furore when, following his reports on the damaging effects of the SSRIs, the University of Toronto withdrew a previously made formal offer of a senior appointment.

Where now?

Many years ago I took part in a conference in the US. The audience was made up of lay people, mainly parents worried about the health and education of their children. My colleagues on the platform were primarily biological psychiatrists, and I listened with fascination and horror as one of them explained how the children's problems arose because of the way in which 'disordered molecules cause diseased minds'. We are currently living through a period in which neuroscience is claiming increasing explanatory power along with the neurotechnology to exploit it. The Human Genome Project offers to lay bare our innermost predispositions and predict our futures thereby. Reductionist medicine and psychiatry seek the explanations for many of our troubles in the presence in our brain of such malfunctioning molecules. The psychiatric casualty load is high and seemingly increasing, with even wider categories of human thought and action becoming subject to the medicalising gaze. The WHO claims that one in five of us is clinically depressed, whilst in the US up to ten per cent of children are said to suffer from learning and attentional disabilities that require regular drug administration. Both the apparent need and the practical possibility of using neuroscience and neurotechnology to manipulate the mind are high on the agenda.

In these contexts, just what is the future of the brain, of human freedom, agency and responsibility? Are we inexorably moving towards a *Brave New World*, the psychotropic paradise envisaged so presciently by Aldous Huxley in the 1930s? In that book, the universal panacea was a mind-changing drug called Soma. 'Hug me till you drug me honey,' runs the book's refrain, 'love's as good as Soma.' The past we record, the present we live, the future we see only as in a glass darkly, but what is clear is that the rapid onward march of technoscience, driven by the demands of the market and the imagination and ingenuity of scientists, is day by day changing the context and fabric of our lives and our societies. In previous decades it was the development of the physical and chemical sciences that provided the motor of change. Today, above all, it is the bio-and info-sciences. Neuroscience, like genetics, lies at the interface of these two, and between them they are bringing hitherto undreamed of questions, prospects and perils into the realm of the possible. Together they are shaping the future of the brain, the future of the persons in whom those brains are embedded, and the future of the societies in which we live. The next chapter begins the exploration of the neurotechnologies and the futures towards which they are directing us via two illuminating case studies.

CHAPTER 10

Modulating the Mind: Mending or Manipulating?

NOVEL BIOTECHNOLOGIES APPEAR ALMOST BY STEALTH. OBSCURE laboratory findings, reported in the arcane language of the scientific journals, may excite the researchers, but seem to be remote from potential application or social concern. Then, rapidly and seemingly without warning, one small further step becomes the subject of patents, of massive research investment, and the public is presented with a *fait accompli*. Research previously disregarded except by enthusiasts as merely pure, disinterested science has become transformed into an imminent technology. Such 'facts on the ground' may be new and seemingly wholly beneficial medical techniques or interventions which rapidly become routine, like MRI to detect damaged brain regions, or the latest generation of pain relievers. Others – the use of foetally derived human stem cells for 'therapeutic cloning', trans-cranial magnetic stimulation, the claims to have located 'genes for' particular traits, or 'non-lethal' chemicals for crowd control – present more complex ethical and social dilemmas; but by the time their existence provokes public unease and scrutiny, it has become too late to consider the options. Once there, the innovation seems irreversible. There is nothing left for 'society' to do but to try to come to terms with the new prospects; to try, however feebly, to regulate their use or legislate their control. Such regulations are what environmentalists sometimes call 'end of pipe' solutions. Or, in the beloved metaphor so redolent of times and technologies long past, it is merely shutting the stable door after the horse has bolted.

Seduced by the claims of the enthusiasts or the promises of profit, or without the political commitment to intervene, successive governmental and international legislation on genetic research and development over the past two decades has almost entirely taken this form – too little, too late, and nearly always reactive rather than proactive. Almost the only example of attempting to block a new technology before it is perfected is the legal prohibition against human reproductive cloning now enacted in many countries – and even that has evoked efforts, notably in the US, to circumvent it.

Whilst the wider public is still (and not unreasonably) focused on the dilemmas raised by the new genetics and reproductive technologies, some powerful techniques for modulating brain and mind are already well established.[1] The prospective advances in knowledge of the brain which the previous chapter considers point the way to new and ever more powerful physical, chemical and biological modes of control and manipulation. Such neurotechnologies are not yet fully formed. Some are under active research, some merely on the drawing board, others nothing but science fiction. There is still time, though not much, for discussion, for consideration of the directions that could, or should, be taken, for rejecting some options and endorsing others – time for civil society and politicians alike to think through the ethical, medical, legal and social issues that the neurotechnologies raise. This task is especially relevant in heavily researching countries like Britain, in the largely unregulated United States and in trans-national entities like the European Union, with its focused research programmes. To point the way towards the future technologies which form the substance of the next chapter, I focus here on two areas of current research, development and use: drugs to enhance memory and cognition, and drugs to control children's behaviour.

'Smart Drugs'[2]

Attempts to enhance human potential and performance are age old: magic potions to produce immortality, superhuman strength, potency and wisdom feature in the myths of most cultures. In the Western tradition they run from the ancient Greeks to the contemporary cartoon characters of Asterix and Obelix. The shelves of health food stores groan with pills offering to improve everything from children's IQ scores to memory in old age. A quick scan of the Internet offers an even wider range of approved and quasi-legal drugs. Many are reportedly available over the counter at a variety of so-called 'smart bars' across the west coast of the US. Pharmaceutical companies race to provide potential treatments for memory loss in Alzheimer's disease, but beyond this already large potential population of users lies a much wider penumbra. The US National Institutes of Mental Health define a category of mild cognitive decline, sometimes called 'age associated memory (or cognitive) impairment'. Some claim that all of us over fifty may be beginning to show signs of such decline and that it becomes much more marked after sixty-five. Hence the interest when, in 1999, a research group led by Joe Tsien, based in Princeton, published a paper in *Nature* reporting that increasing the number of a particular subtype of glutamate receptor in mouse hippocampus by genetic manipulation improved their performance in a test of spatial memory. The paper attracted massive international publicity owing to its provocative conclusion that their findings proved that 'genetic enhancement of mental and cognitive attributes such as intelligence and memory in mammals is possible'.[3] Clearly the potential market for treatments is vast, and, as I mentioned in Chapter 7, over the past decade a number of start-up biotech companies have begun to promise the imminent arrival of drugs that will prevent such an impairment, describing them as 'Viagra for the brain'.[4]

The elision in reports of this sort of memory with cognition is typical. The suggestion that it might be possible to produce drugs with a 'purely' cognitive effect dates back to the 1970s, when Cornelius Giurgia, coined the term 'nootropic' (derived from the Greek *noos* – mind – and *tropein* – to turn) to describe their function. As he put it:

> Do we realise, as individuals or as a species, all of our genetic potential? . . . A pharmacological intervention is more and more feasible and acceptable at all levels of interface between genome and environment. This is the target of the Nootrope endeavour. These drugs, devoid of toxicity or secondary effects, represent a means to enhance plasticity of those neuronal processes directly related to the 'Noosphere' . . . Pharmacology can participate, very modestly, in one of the major efforts of humanity, which is to go beyond the

Platonic question 'Who are we?' . . . Man is not going to wait passively for millions of years before evolution offers him a better brain . . . To develop a pharmacology of integrative action of the brain, in the nootropic sense, seems to me to have a place in this far-reaching human objective.[5]

More prosaically, Dean and Morgenthaler, in a book entitled *Smart Drugs and Nutrients* and subtitled 'How to improve your memory and increase your intelligence using the latest discoveries in neuroscience',[6] argued that:

> The concept of a fixed intelligence is . . . untrue . . . more and more business people and scholars are looking for the kind of 'edge' that athletes get from science . . . Research also shows that you may increase your intelligence by taking certain substances that have recently been shown to improve learning, memory and concentration . . . for improved exam-taking ability, better job performance and increased productivity [as well as] delaying age-related intelligence decrease.

Why enhance cognition?

In an increasingly skills-driven and socially interactive world, memory and cognition are among the keys to success. This is perhaps why loss of memory – an inability to remember – is so mysterious and frightening a condition. As the incidence of disorders like Alzheimer's increases with age, and the age profile of populations in the industrial world is shifting steadily upward, there is a strong medical and social drive for research to develop neuroprotection strategies, or at least to reduce the rate and degree of decline. Most of us share the fear that as we age we will lose our memory, become an Alzheimer patient. 'Solving' AD has become an important target for academic institutions and the pharmaceutical industry. However, in addition to the fear of such relatively unambiguous diseases (although clouded in diagnostic uncertainty until post mortem), many of us fret over our inability to recall names or past events, and are concerned about whether indeed we might be about to suffer from some form of age-associated memory impairment. And then there are all those seeking that competitive 'edge' referred to by Dean and Morgenthaler.

Few would doubt the value of neuroprotection or amelioration of the effects of stroke or AD, but outside this lies the murky area in which 'normality' itself becomes a medical condition. As I've emphasised, some features of cognitive function – notably, speed of processing – seem to decline progressively with age. As one ages, more trials are required to acquire simple conditioned reflexes – but given enough time and trials, the reflex can be acquired. Furthermore, we older people often acquire better strategies for solving problems than our juniors, and our feelings of fading memory might often be associated more with depression.[7] So, targeting 'memory' itself might not be an appropriate strategy even for AD.

The deficits associated with AD and other conditions relate to specific biochemical or physiological lesions. There is therefore no *a priori* reason – irrespective of ethical concerns or any other arguments – to suppose that, in the absence of pathology, pharmacological enhancement of these biochemical processes will necessarily enhance memory or cognition, which might already be 'set' at psychologically optimal levels. If they are suboptimal, this might reflect not a pharmacological deficit, but other social or personal life-history reasons. This is not to minimise the distress that most of us feel in our daily life as a consequence of lapses of memory, but, as politicians, card sharps and competitors for the *Guinness Book of Records* know, pharmacological intervention is not the only route to overcoming such problems. Memory – for names, hands of cards dealt, or even recalling π to hundreds of decimal places – can be trained using non-pharmacological techniques that date back to antiquity.[8]

Furthermore, it is worth querying the assumption that a perfect long-term memory is desirable. The psychological mechanisms of perceptual filtering, and of short-term, recognition and working memory, are clearly beneficial in blocking the accumulation of irrelevant or transiently required information in longer-term stores. The wisdom of the 'recovery' of past traumatic memories by psychotherapy or psychoanalysis has been questioned, and even the veracity of such apparent memories has been challenged in the context of so-called 'false memory syndrome'.[9] Therapeutic forgetting might indeed be beneficial.

The literature is full of anecdotal accounts of the problems faced by those few people who are apparently unable to use forgetting mechanisms to discard unwanted information, and to assimilate only the necessary. The most famous case is that of Shereshevkii, the patient who was studied over many years by the neuropsychologist Alexander Luria.[10] Shereshevkii had an apparently inexhaustible memory, recalling not merely complex nonsense formulae, but also the exact context in which he learnt them. His inability to forget made it impossible for him to hold down a career other than as a memory performer. His case poignantly echoes that of Funes, the 'memorious' – the fictional character created by the novelist Jorge Luis Borges – who had '. . . more memories in myself alone than all men have had since the world was a world . . . my memory, sir, is like a garbage disposal . . .'[11] It is no surprise that, in the story, Funes dies young – of an overdose of memory, so to speak.

Nootropes, remembering and forgetting

The implication of the nootrope concept is that there are processes in the brain that are purely concerned with memory, remembering and forgetting, and that drugs can be developed to affect these processes without peripheral or other central effects. Both propositions are questionable. Both learning and remembering require, among other mental processes, perception, attention and arousal, which involve not just brain but body processes. So, agents that affect any of these processes might also function to enhance (or inhibit) cognitive performance.

In both humans and other animals, learning and remembering are affected by the levels of circulating steroids in the bloods, by adrenaline, and even by blood-sugar levels.[12] Central processes can also affect performance by reducing anxiety, enhancing attention or increasing the salience of the experience to be learnt and remembered. Amphetamines, methylphenidate (Ritalin), antidepressants and anxiolytics probably act in this way. Other hormones that are regularly cited as potential smart drugs, such as adrenocorticotropic hormone (ACTH) and vasopressin,[13] might function in a similar fashion. Steroid hormones, such as oestrogen, and neurosteroids, such as dehydroepiandrosterone (DHEA), and growth factors, such as BDNF, will also enhance memory in animal experiments, though clinical trials with oestrogen (as in hormone replacement therapy) have not proved encouraging.[14]

Approaches to enhancement

Visit any health food stores or check out the Internet for cognition and memory improvers and you will find a long list of substances such as lecithin and multivitamins – notably, the B complex and vitamin C – along with herbal extracts of ginseng, ginkgo biloba and other substances derived from non-Western, non-allopathic traditions. More allopathic approaches to enhancement have tended to follow clinical fashion in identifying the physiological or biochemical processes that are impaired in cognitive deficit, and focusing on ameliorating them. Thus, suggestions that one of the main problems of cognition in ageing lay in deficits in general cerebral metabolism provided the impetus for nootropics that were supposed to boost the circulation and oxygen use. An example is co-dergocrine mesilate (Hydergine), an anti-hypertensive fungal extract that is claimed by Dean and Morgenthaler to 'increase intelligence, memory, learning and recall', among a dazzling array of other virtues. The *British National Formulary*, by contrast, states that 'the drugs have not been shown clinically to be of much benefit'. Some drugs that affect calcium uptake into neurons also enhance memory formation in experimental animals, but once more clinical trials in humans have not proved effective.

The cholinergic hypothesis for the memory deficits associated with AD led to an intensive search for drugs that might restore cholinergic function – three of the currently licensed drugs work like this (see Chapter 7). Again, animal models pointed the way. A number of experiments suggested that if an animal is trained on some task and a substance that blocks acetylcholine neurotransmission (for instance, the drug scopolamine) is injected, the animal becomes amnesic and forgets the training. Drugs that restore cholinergic function can protect against such amnesia. However, most have proved to be ineffective at ameliorating the deficits of AD in humans, still less as general memory or cognition enhancers. This is not really surprising, as the logic of the experiments is essentially circular: scopolamine produces learning deficits, so agents that counteract scopolamine prevent these deficits.

However, unless the memory deficit in humans is indeed caused by a scopolamine-like blockade of cholinergic function, it is not likely to respond in the same way. As all the cholinergic drugs can produce unpleasant adverse reactions and are only mildly efficacious even in AD, it is hard to imagine their more general use as treatments for age-associated memory impairment (or as nootropics in the Giurgea sense), although they are apparently under trial.

Another potential route for memory enhancement comes via inter-actions with glutamate neurotransmission. The experiment by Tsien and his group showed that some types of memory can be improved by increasing the amount of a particular class of glutamate receptor. Drugs that interact with another type of glutamate receptor (ampakines) are also said to enhance memory retention and are currently in clinical trial. It was the ampakines which were allegedly promoted at a press confer-ence as giving 'a seventy-year-old the memory of a twenty-year-old'. But glutamate receptors were also the target for earlier generations of alleged smart drugs, a family of chemicals called acetams, still available over the Internet and once lauded as 'pure nootropics' though of dubious effec-tiveness in clinical trials.

Molecular biologists studying the biochemical cascade that follows transmitter release in animal models of learning have identified a sequence of steps that leads ultimately to increased gene expression and the synthesis of new proteins. A key step, found in both fruitflies and mice, involves a protein rejoicing in the name cyclic-AMP-responsive element-binding protein (or CREB).[15] At least two companies (those referred to in the *Forbes* article on 'Viagra for the Brain' discussed on page 245) have been set up to explore its potential, and related prod-ucts are already in clinical trial, even though in animal models, the role of CREB in retention seems to depend rather sensitively on the training protocols that are used.[16]

This illustrates a more general issue: the relevance of animal models in the field of memory and cognition. It is striking that, despite clearcut evidence that a variety of substances can enhance memory-related performance in animals, they have generally proved to be disappointing in treating cognitive decline and dementia when taken to clinical trial. There are several possible reasons for this. Alzheimer's is a specific disease, and the biochemical processes that it involves cannot easily be modelled in animals. More importantly, in animals learning and recall must always be tested by the criterion of performance of some task, such as remembering the way through a maze. The analogy with the subtleties of human verbal, recognition and autobiographical memory may not extend fully to the biochemical mechanisms that memory engages. General 'cognition' is hard to test in animal models (except perhaps in complex tasks with primates), and memory is but one aspect of cognition in humans. This is not to deny the utility of animal studies – far from it, or I would long ago have ceased working with my chicks – but it is to recognise their limitations.

Do we want cognitive enhancement?

It might seem self-evident that protection against cognitive impairment and the recovery of cognitive functions in the absence of proactive treatment, if possible, are desirable, but to 'give a seventy-year-old the memory of a twenty-year-old' requires a little more discussion before nodding and passing on to the wider issues. Memory loss confounds at least two distinct phenomena. To most people, it implies loss of long-term autobiographical memory, arguably the feature of patients with AD that is most distressing to carers. Drug treatment is conceived of as aiding in the recovery of these lost memories, but there is no indication that any of the agents under discussion as putative cognitive enhancers, or for therapeutic intervention in AD, will achieve this. Rather, they might serve to prevent loss of short-term memory for recent events – that is, to aid in the transition between short-and long-term memory. As forgetfulness for recent events (Did I do the shopping? Where did I leave my keys?) is one of the characteristic features of the earlier stages of AD, the newer generation of enhancers, including those that reverse some of the specific biochemical lesions, could function to alleviate these early features, enabling those suffering from AD to remain independent for longer. It is improbable that they will reverse or prevent the progression of the disease. Indeed, as memory loss is only one feature of the disease, treating it alone may not be desirable. One recent National Health Service study suggested that the use of Aricept and related drugs produced almost no measurable improvement in quality of life for people suffering from Alzheimer's.[17] And, even if a really effective agent could be developed that did awaken long-dormant or even erased memories in patients in the advanced stages of the disease, such memories might not necessarily be welcome. These re-awakenings might be as painful as those documented by Oliver Sacks in his famous account of how l-dopa temporarily brought back to consciousness people who had been in a narcoleptic state for decades.[18]

Neuroprotection would seem to be a better strategy, but, once again, it is not an unequivocal good. Some of the genetic and environmental risk factors for AD are understood, but almost all are at best only weakly predictive. One would have to weigh carefully the risks and costs of long-term medication, and the penumbra of its perhaps inappropriate use by the so-called 'worried well', concerned about failing memory, and the peculiar diagnosis of age-associated memory impairment. As a society, we are becoming familiar with long-term preventive medication – for example, with the use of anti-hypertensives and statins to reduce the risk of coronary heart disease for those who are judged vulnerable. However, the assessment of risk factors and weighing of inevitable adverse and unwanted drug effects are tricky even in physiological and biochemical contexts that are better understood than cognition. As I keep emphasising, growing old is not a disease but a condition of life, and one can question the consequences of creating a society that refuses to accept ageing – at least for the wealthy.

Beyond these potential clinical and neuroprotective uses for the cognitive enhancers is the terrain that raises most ethical and legal concern – their potential for improving, as Dean and Morgenthaler put it, school and examination performance and 'competitive edge'. During their missions in the 2003 Iraq war, US pilots were reported to have been fed on a variety of drugs to keep them awake, so it is perhaps not surprising that the US airforce is researching the effects on pilot performance of one of the cholinergic drugs (donepezil) licensed for treatment of AD.[19] But is such enhancement theoretically possible? Despite the problems that those of us with weak memories experience in our daily life, more does not necessarily mean better. So, even if a pharmacological remedy for deficits such as those in AD were developed, this would not automatically mean that a supernormal level of the relevant molecule would produce supernormal performance. Where brain processes depend on a subtle balance between neuromodulators, neurotransmitters and their multiple receptors, simply adding more of one (such as a specific glutamate receptor) might be more disruptive than beneficial. Even if this proved not to be the case, and safe and effective enhancers of 'normality' could be produced, there is a fine medical and ethical line between correcting deficits and improving on 'normality'.

The issues are analogous to those raised by the uses of steroids and other performance enhancers in athletics, where a sort of arms race has developed between the athletes who might use them, and the rule-makers and enforcement systems that detect and ban them.

We should not be naïve, however. Generations of students (to say nothing of creative artists or frenetic dealers in stock markets) have used such stimulants as have been available – caffeine, alcohol, amphetamines – to sharpen concentration as they revise for and sit examinations. Would the availability of genuinely effective new drugs make any difference in principle?

Perhaps not, but one can foresee interesting legal issues arising if the losers in some competitive examination cry foul and seek redress.[20] The clutch of insurance and related cases that surround the use of Prozac and other SSRIs, especially in the United States, are a foretaste of what might arise in this new context. As I've said, social thinking and policy on the uses of chemicals that affect brain or body performance are hopelessly confused about existing substances. Within the foreseeable future, cognitive enhancers – or agents that are claimed to function as cognitive enhancers, whether or not they are genuinely effective – will join this eclectic set.[21] My best guess is that, as with steroids for athletes, they will turn out to be virtually uncontrollable legally, and, as a society, we are going to have to learn to live with them. I will return at the end of this chapter to the implications.

Ritalin

In 1902, the paediatrician George Still, writing in the *Lancet*, described a child patient as 'passionate, deviant, spiteful, and lacking in inhibitory volition'.[22] But it wasn't until 1968 that the American Psychiatric Association (APA) standardised a set of criteria for what was then called 'hyperkinetic reaction of childhood'. At that time, the gulf between the two sides of the Atlantic seemed very wide. In Britain, troublesome youngsters tended to be classified as disturbed or naughty and, if unmanageable in class, were placed in special schools or segregated units. There, children might be treated by behaviour modification therapies, so-called 'token economies' with loss of privileges for poor behaviour and rewards for being good. Such children are nowadays classified as suffering from 'Emotional, Behavioural and Social Difficulties' (EBSD), caused by defective socialisation: for example, lack of parental control, family break-up, abuse or deprivation. Their parents are supposed to receive support from social workers, though in reality this may be pretty minimal. As for hyperkinesis, for several post-1968 decades British paediatricians would continue to say that although there were cases associated with specific brain damage, these were very rare, affecting perhaps one in a thousand children.

Something very different was taking place in the US. There, it seemed that, even in the 1960s, between 3 and 7 per cent of young children, nine times more boys than girls, were suffering from a specific disease as classified by the APA. Over the subsequent decades the name of the disease changed. To start with it was Minimal Brain Damage, but when no such overt damage could be detected it became Minimal Brain Dysfunction, then Attention Deficit Disorder, and, most recently Attention Deficit/Hyperactivity Disorder (ADHD). The latest edition of the *US Diagnostic and Statistical Manual* characterises a child (or adult) suffering from ADHD as being 'severely inattentive (forgetful, careless, etc.) or hyperactive/impulsive (restless, impatient, aggressive, etc.)' for at least six months. The symptoms are said to emerge by the age of seven, and produce problems for the child at home and at school.[23] Children untreated for ADHD are said to run a greater risk of becoming criminals as they grow older.

Now the difficulty with these diagnostic criteria, it should be immediately apparent, is that the description of the behavioural problem is of its nature relational, in that it requires comparison of the child being diagnosed with other children of the same age, background and so on. Is the ADHD child more inattentive or restless than others in his group? That is, there needs to be a definition of normality against which the child is to be judged abnormal – and the narrower that range, the more children fall outside it. Almost certainly, the range of what is considered normal in British schools is broader than in the US – at least for white British children, because there has for many years been a grossly disproportionate exclusion of children from black British families.

However, it is at this point that an interesting elision occurs, because the word 'normal' has two meanings that readily become confounded.

There is 'normality' in the statistical sense. This relates to the famous (or infamous) bell curve pattern of distribution of some variable around the mean. A value of the variable – say, some measure of behaviour, classically IQ – is then called 'normal' if it falls within two standard deviations of the mean. But there is also the more everyday use of the word normal, the normative sense that implies a value judgement – essentially subjective, or socially determined – of how a child should, or is expected to, behave. A child who does not behave 'normally' in this sense is clearly 'abnormal'.

So what are the typical behaviours that lead to a child being classified as suffering from ADHD? As the educationalist Paul Cooper describes them,

> children with AD/HD are often portrayed as being of average to high ability, but they disturb their parents and teachers because their classroom achievement is erratic, and often below their apparent level of ability. The child with AD/HD will often be a source of exasperation to the generally effective teacher. On occasions the child may show high levels of performance, a ready wit and an imagination of a high order but with erratic performance. In one form of the disorder he or she may often appear disengaged, easily distracted and lacking in motivation. The child may appear to be lazy or wasting obvious ability in favour of being oppositional and disruptive. This is the pupil who never seems to be in his or her seat, who is constantly bothering classmates, and can be relied upon for little other than being generally off task. All categories can be frustrating to teach because of their apparent unpredictability; their failure to conform to expectations, and their tendency not to learn from their mistakes.[24]

This description makes clear the relational sense in which the child is being discussed – his behaviour is being defined in relation to others: teacher, fellow students, parents. However, the ADHD child is not simply naughty or ill-disciplined. The origins of his aberrant behaviour do not lie in his interactions with his teacher and the rest of his class, or with his parents, his home circumstances, his teacher's adequacy or the school

environment – all those considerations that underlie the EBSD classification. The 'cause' of ADHD is now unequivocally located inside the
child's head. Something is wrong with his brain. This has led critics to
describe ADHD not as a disorder but as a cultural construct, in a society
that seeks to relocate problems from the social to the individual.[25]

Just what might be wrong with such children's brains is unclear, as
there are no manifest structural or biochemical abnormalities in those
diagnosed with ADHD. There are some neuroimaging studies, but these
are not properly controlled and produce conflicting results. However,
the consensus amongst those who are convinced of the legitimacy of
the diagnosis is that the 'something' is a fault in dopamine neurotransmission, caused by some genetic predisposition. The genetic case
is based on initial observations, made in the 1970s, that ADHD 'tends
to run in families'. A comparison was made of fifty children diagnosed
as hyperactive in a US hospital outpatients department, compared with
fifty matched controls referred for surgery. Parents were interviewed
about themselves and their children's behaviour and family history. Nine
of the 'controls' were also said by their parents to be 'wild' or 'out of
control' and transferred to the hyperactive group. The parents of the
hyperactives had a high incidence of alcoholism, 'sociopathy', etcetera,
and the researchers – who knew in advance which were the 'hyperactive' and which the 'control' families – quizzed the parents about their
own relatives. They concluded that hyperactivity ran in the family and
was likely to be inherited.[26] It is clear that this methodology (repeated
soon by other US researchers) is deeply flawed.[27]

There followed the seemingly obligatory twin studies, in which identical (monozygotic, MZ) twins are compared with non-identical, dizy-
gotic (DZ) twins. These suggest that if one MZ twin is diagnosed ADHD,
the other has an above average risk of the same diagnosis. That is, the
twins are concordant. DZ twins show lower concordance. The rationale
for such twin studies seems straightforward. Identical twins share 100
per cent of their alleles, whereas DZ twins like other siblings share only
50 per cent. Yet as MZ and DZ twins also normally share common
environments, any increase in concordance in the MZs over the DZs
must be due to their genes. An alternative version of the twin method
works by identifying the few cases in which MZ twins have been separated at birth and put into different families for adoption. In this
situation, the argument goes, individuals with identical genotypes are

reared in different environments, and the behaviour of each can be compared both with that of the other twin and of any non-genetically related children in the adoptive household. These 'twin methods' and the associated calculation of a measure called 'heritability' has been a staple of behavioural psychology for decades now. Such estimates do not, as is sometimes supposed, answer the question of how much genes and environment contribute to any individual's behaviour; as everyone recognises, this is a strictly meaningless question. What they do claim to do is to partition out the *variance* between individuals in a population between genes and environment. A heritability of 100 per cent should indicate that all the differences between people are genetic in origin, and of 0 per cent that all are environmental. The estimate depends in the first place on being able to define and preferably quantify the behaviour or attribute being studied (intelligence, aggression, school behaviour or whatever). It also depends on there being a uniform environment across which genotypes are distributed; if the environment changes the heritability estimate changes too. Thus if it were possible to imagine a situation in which all children grew up in precisely identical environments in all senses of that word, then all the differences between them would be due to genes, and heritability would be 100 per cent! This is why heritability is also a 'within population' measure; it can't be used to compare populations who are not randomly interbreeding (for example, blacks and whites in the US, working and middle classes in the UK).

These caveats haven't prevented heritability estimates being used to suggest that a wide variety of human behaviours and beliefs are, at least in part, determined genetically. The use of twin studies and heritability estimates has been heavily criticised, by me amongst others, for their general methodological assumptions, and I don't want to re-enter that territory now as my agenda lies in another direction.[28] Suffice it to say that there are many reasons other than identical genes why MZ twins might behave more similarly and be treated more similarly than DZs, many reasons why the ideal model of 'twins reared apart' rarely obtains in practice. For most purposes, twin research has been superseded by molecular methods that focus specifically on identifying specific genes or gene markers, and only a small group of behaviour geneticists clings almost religiously to this obsolete technique[29].

More modern genetic methods have been employed in the context of ADHD, and their use has led to claims that at least some of the genes associated with the proteins involved in dopamine neurotransmission might be abnormal in the disorder, but even here the evidence is elusive. There is no single gene that might be involved, rather, it is claimed, a number of risk factors, which, in the context of precipitating environmental factors, such as the types of social deprivation or family breakdown mentioned above, might result in the behaviour which would lead to the ADHD diagnosis.[30]

Because a child diagnosed with ADHD isn't seen as being naughty or deprived, he neither needs to be punished nor offered social support. Instead, he becomes a medical problem, and, from the 1960s on, children diagnosed with minimal brain dysfunction or ADHD in the US began to be treated with a drug that enhances dopamine neurotransmission. This is methylphenidate (Ritalin), patented by Ciba-Geigy, (now Novartis). Methylphenidate's effects are rather similar to those of amphetamine, which was widely used during the 1960s as a stimulant. It is an interesting reflection on the change in attitude to psychoactive drugs over time that back then amphetamine was widely known as 'speed', and viewed with suspicion, rather as Ecstasy is now. So widespread did its use become that in the UK doctors were only allowed to prescribe it under licence. When methylphenidate began to be prescribed in the US, it was claimed that instead of acting as a stimulant, it had a calming effect on ADHD children, improving focus and attention, reducing disruptive behaviour and benefiting school performance. At first, this was regarded as paradoxical: why should a drug called 'speed' have a calming effect on the children? Advocates of the diagnosis and the use of Ritalin embarked on convoluted arguments as to why this should be the case, and even suggested that ADHD could be diagnosed on the basis of whether Ritalin had this paradoxical effect – another neat example of *ex juvantibus* logic. Just as depression might be the result of too little serotonin in the brain, the argument went, ADHD might be the result of too little dopamine.

However, before long it became apparent that the drug affected 'normal' and 'ADHD' children similarly; there was in fact no paradox. So the neurobiological argument shifted. It was proposed that by enhancing dopamine neurotransmission, methylphenidate 'improved communication' between frontal cortex and midbrain/limbic system. This, it was hypothesised, put the 'higher' brain region in control of the 'lower', more impulsive brain systems, and hence diminished the excesses of the hyperactive child. It has to be said that there is no serious evidence to support this elaborate conjecture, but, just as Prozac made 'normal' people 'better than well', so, it was argued, Ritalin might enhance school performance even in 'normal' children. For the important thing is that the drug seemed to work. Children on Ritalin are said to become less fidgety in class and to show an improvement in 'behavior which is perceived by teachers as disruptive and socially inappropriate'.[31] No wonder that ADHD soon began to be diagnosed on the basis of school-teachers' reports and was observed to show a peculiar expression pattern, often remitting at weekends and in school holidays.

As to how the children themselves experience taking the drug, Cooper quotes a number of revealing comments amongst British school students:

> When I'm on it [Ritalin] I work harder, and I'm nicer, but when I'm out of school [and not on Ritalin] I'm sometimes silly, or I act stupid, or do things that I wouldn't really do if I was on Ritalin . . . When I'm on [Ritalin] I have more control over what I say . . . (girl, 12)

> When I'm taking Ritalin I'm calmer. I can study more and everything. And when I'm not I really can't concentrate or anything (girl, 13)

> I can concentrate better on Ritalin, I think like. I get on with my work more, and I don't talk so much (boy, 14)

> It makes me – Ritalin and Pemoline and things – they make me think first. I can think for myself anyway, but they make me think even better for myself (boy, 15)

Others were more negative, claiming that it 'mucked up your head'.

> Sometimes I like it [Ritalin], but sometimes I don't . . . If I do take it when we didn't have school, I wouldn't want to go outside and play with my friends, or, I would just want to stay in home by myself and read a book or watch television or something (girl, 15)

The clear implication is that, whatever it does for the child, Ritalin makes life more comfortable for harassed teachers and parents. The result has been that the pressure to prescribe the drug, both in the US and more recently in the UK, has come largely from outside the medical and psychiatric profession, and often from parents themselves, especially during the 1990s which in the US was a period of increasing class sizes, rejection of physical punishment in schools, and special funding for schools that identified children with disabilities like ADHD.[32] The growth in Ritalin prescriptions in the US has been dramatic, from a few hundred thousand in the 1980s to as many as eight million today. Between 1989 and 1997 alone the number of US children diagnosed as having 'learning disorders' increased by 37 per cent to 2.6 million, while over the same period the number of ADHD diagnoses increased nine-fold. The alarm bells were being rung by many,[33] but it wasn't until the end of the decade that the Food and Drug Administration in the US and the WHO internationally began to warn of an 'epidemic' of drug taking, with Ritalin being traded in school playgrounds along with other less legal substances.

Meanwhile, a variety of other drugs was being added to the torrent now being enthusiastically and quite legally poured down children's throats. In 1997 the UK-based drug company Shire bought the rights to a failed juvenile obesity drug that researchers had noticed had a calming effect on children, and 'repositioned' the substance dexamphetamine (Adderall) as a treatment for ADHD.[34] Ritalin has to be taken through the day (and is apparently sometimes dispensed by teachers themselves), whereas Adderall is a slow-release formulation that needs be taken only before and after school and is therefore less likely to be abused. This 'repositioning' is interesting. The history of the introduction of psychotropic drugs is frequently one in which a drug originally developed with one potential purpose in mind finds itself used for another. This can sometimes be benign serendipity, as when quinine, used in the treatment of malaria, also finds a use in relieving muscle cramps. However, as the psychiatrist and critic of many of the claims of the pharmaceutical industry David Healy points out, it is also often part of a deliberate strategy of the industry in relocating their products in the context of changing fashions in psychiatric diagnosis.[35]

So far as Adderall is concerned, in his interview with Joe Studwell published in the *Financial Times*, Shire's chief financial officer, Angus

Russell, was commendably frank. Shire gathered all available data on the 180,000 US-based clinical psychiatrists, paediatricians and general practitioners who had prescribed an attention deficit drug and identified 'a subgroup of 27,000 doctors who had written 80 per cent of ADHD scrips. Within this group tiers of American physicians were selected according to the volume of drugs they prescribed. A sales strategy was mapped out accordingly. "We literally treat it as a pyramid," says Russell. "The first 1000 physicians probably prescribe 15 per cent of the market. The top 1000 get 35 visits a year."' Shire now has 23 per cent of the ADHD market in the US, earning $250million profit out of revenues of just over $1billion. This is hard-selling big business.

Adderall, as an amphetamine-like substance, is currently unlikely to be readily licensed in the UK. However, in June 2004, a further drug, Strattera, produced by Eli Lilly, was cleared for prescription. Strattera doesn't interact with dopamine receptors at all, but is instead a specific reuptake inhibitor for quite another neurotransmitter, noradrenaline. Like Adderall it is said to have the advantage of requiring only a single daily dose, thus making it unnecessary for the child to be given another pill whilst at school, and therefore better managed and supervised by parents – and indeed, as the Strattera publicity hints, therefore less likely to be 'traded' by children prescribed it. What makes the whole ADHD story even less credible, however, is the fact that until now the claimed brain 'lesion' in the condition has been in the dopamine system, and yet now a drug with an entirely different target site is also said to work.

As I've said, for a long time there seemed to be an extraordinary difference between the behaviour of children in the US and the UK. British child psychologists were diagnosing only at one-tenth the rate found amongst young Americans, and Ritalin was not being prescribed. Why? If the disease is heritable, what distinguishes the US from the UK genotype? Alternatively, is there something especially disease-provoking in the way that US children are reared? If neither of these explanations is acceptable, then there are only two possibilities. Either, there is a huge level of over-diagnosis in the US, partly as a result of teacher and parent pressure, or – a much more attractive proposition to proponents of ADHD – the British are simply behind the times in their reluctance to diagnose a childhood disease.

In any event, the UK, along with other European countries such as Sweden and Germany, is now catching up fast, although prescription rates are still much lower than in the US or Australia, with ADHD incidence rates quoted, even by the most gung-ho enthusiast, of around 1 per cent – a third to a fifth of the US prevalence. In the early 1990s, when prescription of amphetamine-like drugs was still fairly rigorously controlled in the aftermath of the panic about addiction which resulted from the ready availability of the drug in the 1960s, Ritalin prescriptions were running at about 2000 a year. By 1997 the prescription level had increased nearly fifty-fold, to 92,000 a year and by 2002 the figure was around 150,000. In Scotland, for instance, prescriptions increased by 68 per cent between 1999 and 2003. There is no sign yet of the rise levelling off.

Why this huge shift over the decade? Although, unlike the situation in the US, pharmaceutical companies are not allowed to advertise directly to the public, pressure to 'recognise' the disorder and to prescribe Ritalin seems to have come initially from parents themselves, sometimes in collaboration with doctors or child psychiatrists working privately, outside the NHS. At least one powerful parents' group was established in Birmingham by a father who had been in the US and had become convinced that both he and his son had ADHD. Local associations began to be formed, to pressurise for a change in prescribing practices. A few powerfully-placed child psychiatrists with a strong biological rather than social bent also began bringing the diagnosis and the prescribing practices into their routine. Clear evidence of abuse began to emerge, with children as young as two being placed on the drug. I have had reports from concerned parents about their children being placed on cocktails of antidepressants, Ritalin and anti-epileptics, despite the clear warnings given in the *National Formulary*.

The media became interested, with television programmes portraying parents trying to control their hyperactive children and singing the praises of the drug, and the Ritalin bandwagon began to roll. Both the British Psychological Society and the National Institute for Clinical Excellence (NICE), which evaluates new drugs and procedures, pronounced themselves cautiously in favour of Ritalin (and NICE has now approved Strattera), although it was only to be used in controlled circumstances and with associated special needs care for the child.

Movements generate counter-movements. Other parents' groups opposed to the ADHD diagnosis, or at the least to the prescription of Ritalin in its treatment, began to emerge – one of the most active is currently based in Edinburgh.[36] Some at least of these oppositional groups do not dispute that children are hyperactive, but instead seek for explanations based either on consumption of junk food or unhealthy food additives, or television or computer addiction. In some ways the debate runs parallel to that beginning to emerge in Britain over the alarming rise in childhood obesity and its possible causes and treatments. There is a curious symmetry between the arguments of parental advocates of Ritalin and of diet control for ADHD. Both attribute the proximal cause of the child's behaviour to some physiological problem to be treated by altering the pattern of consumption, by adding a drug or removing allegedly unhealthy foods. So both the 'genetic' and the 'environmental' advocates downplay the societal, relational issues and instead seek for individual solutions to what others would regard as in large measure social problems. Both, though each would deny it, are displaying a form of biological determinism.

Unlike the present generation of cognitive enhancers, there is no doubt that Ritalin 'works', as in the testimony of the children interviewed by Cooper quoted above, though none of these is perhaps as poignant as the American child quoted in the pro-Ritalin literature who spoke of his 'magic pills which make me quiet and everybody like me'. Children who are calmer at home and school are easier to parent and to teach. However, Ritalin no more 'cures' ADHD than aspirin cures toothache. Masking the psychic pain that disruptive behaviour indicates can provide a breathing space for parents, teachers and the child to negotiate a new and better relationship, but if the opportunity to do this is not seized, we will once again find ourselves trying to adjust the mind rather than adjust society. Just how powerful the new techniques aimed at doing just this are likely to be is the subject of the next chapter.

CHAPTER 11

The Next Big Thing?

COGNITIVE ENHANCERS AND RITALIN EXEMPLIFY TWO IMPORTANT FEATURES of the psychocivilised society into which we are moving: on the one hand its essential individualism, on the other, increasingly sophisticated methods of control and seemingly non-violent coercion. In the context of the historical relationship between genetics and eugenics, Hilary Rose has documented the transition between the earlier phases of eugenic policy, with its state-regulated and fostered programmes of compulsory sterilisation (and of course the Nazis' Final Solution), and the present day, when the variety of new reproductive technologies and the ideology of individual choice opens up instead the prospect of what she calls 'consumer eugenics'.[1] Thus too with the new generation of psycho-chemicals; it will be an individual's choice whether and when to take a cognitive enhancer or a designer mood-changer. However, it may not be left to individual parents whether and when to drug their child; this may be imposed by the educational authorities or new variants of state psycho-social services. Coupled with the licit and illicit production of newer and more sophisticated mood enhancers and happiness pills to make one 'better than well', the future offers the prospect of an entire population drifting through life in a drug-induced haze of contentment, no longer dissatisfied with their life prospects, or with the more global prospects of society in general, with the neurotechnology to erase such tremors of dissent that may remain added to the already formidable existing armoury of state control.

But of that more later; the first task is to identify the potential technologies themselves. A convenient check list is provided by the September 2003 special issue of *Scientific American* entitled 'Better Brains: how neuroscience will enhance you'. Its cover summed up the themes it intended to cover: 'Ultimate self-improvement; New hope for brain repair; The quest for a smart pill; Brain stimulators; Taming stress; Mind-reading machines; Genes of the psyche' and, inevitably, last of all, 'Neuroethics'. The US President's Council on Bioethics had a slightly different list: 'The pursuit of happiness; better children; superior performance; ageless bodies; happy souls.'[2] So what lies behind such headlines? In this chapter I will try to ground some of these future prospects by considering their current status. In doing so I will focus on developments that seem to me to give rise to ethical concern. There are surely enough amongst my neuroscientific colleagues to point to the potential beneficent neurotechnological prospects of the coming decades. These include overcoming the refusal of adult nerve cells to regenerate so as to be able to treat spinal injuries or autoimmune diseases like multiple sclerosis, or even brain damage itself; better drugs to relieve the 'black dog' of depression or the agonies of schizophrenia; gene-based therapies for Huntington's disease and other neurological disorders. But the triumphalist tone in which such advances are foreshadowed leaves a residue of worries about the extent to which these new technologies open the way to mind manipulation, to the restriction of our concepts of human agency and responsibility, to the creation of a psychocivilised society that has indeed moved beyond freedom and dignity. These, above all, are the concerns that I wish to highlight in this chapter. There seem to me to be two broad issues towards which the technologies are directed. The first is that of predicting possible future behaviours or present intentions from genetic, neurochemical or neuroimaging data. The second is that of changing or directing that behaviour by direct intervention. It is true that, as I shall argue, many of the proposed technologies lie in the realm of science fiction, whilst others resemble the lurid promises of snake oil salesmen. However, technologies may have powerful consequences even when based on dubious science, as the Ritalin example shows, and we cannot afford to ignore them, even when doubting their rationale.

Reading the mind

It may not be necessary to go inside the brain to read minds. One of Charles Darwin's less well-known books is entitled *The Expression of the Emotions in Man and Animals*. In it Darwin suggests that there are a number of fundamental human emotions, amongst them anger, fear, joy and disgust, which are expressed in the face in ways which are cross-culturally and universally recognisable. He illustrates this with photographs of what seem to a modern eye remarkably ham Victorian actors apparently expressing these emotions. Later generations of anthropologists have built upon this claim; these four basic emotions, they argue, are expressed through similar facial expressions whether in Japan, New Guinea or the US. Indeed, such universalism forms one of the stronger claims of evolutionary psychology. Cultural, class, gender, religious and other social forces can modify these expressions, it is admitted, but cannot entirely eliminate them. One of the leading exponents of such face-reading, Paul Ekman,[3] has developed techniques which have been used by the US police and intelligence agencies in the context of interrogation procedures. He argues that it is possible to detect brief micro-movements of facial muscles on which cultural constraints are subsequently imposed, and to distinguish between 'genuine' and 'simulated' emotions accordingly. I have to confess that I must be singularly inefficient at reading such expressions, judging by my failures to interpret the photos Ekman employs, which seem to me to be simply more modern versions of Darwin's hammy acting. However, if he is right, and his methods can be taught, then the days of (face-to-face, as opposed to Internet) poker playing are numbered. One can no longer take the possibility of the privacy of one's thoughts and emotions for granted. Can psychocivilisation go further?

But Ekman's are only surface manifestations of neural processes. If one could use fMRI or MEG to pick up some particular pattern of brain activity in response to specific stimuli, then one might be able not only to read present emotions but to predict subsequent behaviour. As I've made clear, such imaging techniques, though powerful, are still in their infancy, though attracting more and more interest and associated funding. Some of the problems with interpreting and predicting from brain images are obvious. Sure, they are excellent at identifying specific areas of damage or disease, changes in volume of particular structures and so forth, but for more subtle variations there comes the problem of comparison, of defining normality, of comparing person X's brain image with some standard, and of interpreting the direction of causation. For instance, in the case of the imaging of the brains of children diagnosed with ADHD the problem is that the children concerned have been medicated, so there is no way of telling whether the claimed differences from the 'normal' are the consequence of the medication or the 'cause' of the behaviour which has led to the children being medicated. Or indeed, whether the 'behaviour' itself has 'caused' the changes in the image. Correlations are important, but they cannot in themselves indicate causation.

However, the data coming from the imagers is becoming more and more intriguing. It is not only a question of being able to find a spot in the brain that lights up when a religious person thinks of God, or meditates on the sublime, or declares themself to be in love. This all seems innocent enough. But how about the claims that brain imaging can reveal racial prejudice, apparently revealed by changes in a specific cortical region, the fusiform cortex[4], or that interracial contact affects the brain's 'executive function'?[5] Or to be able to tell from brain imaging whether a person is consciously lying, or to distinguish 'false' from 'true' memories?

Could one envisage a world in which neuroimaging becomes more than a research and clinical tool and instead one providing practical guidance in commerce and law? Certainly the prospect is more than a glint in the eye. Ever since we published our MEG study of supermarket shopping, discussed in Chapter 6, we have been besieged with invitations to marketing conferences intended to interest companies in using imaging techniques to improve their appeal to customers – so called neuromarketing (or, sometimes, neuroeconomics). Car manufacturers seem particularly keen to exploit the new techniques, and both Ford and DaimlerChrysler have begun to use scanning to check the impact of their products. DaimlerChrysler has even established a lab (MindLab) in Ulm, Germany, using fMRI, to carry forward the work, whilst in the US researchers are investigating the neural processes involved in choosing between Coca-Cola and Pepsi-Cola.[6]

It is of course not merely commerce that has expressed enthusiasm. There's been a long interest by military intelligence in the prospects and the possibilities of devising technologies for mind-reading and thought control. Research initially focused on using EEG and MEG to identify neural processes associated with word recognition, and to differentiate brain signals distinguishing words from pictures and nouns from verbs.[7] Solzhenitsyn's *First Circle* describes a primitive version of such experiments, based on voice-prints, conducted by psychologists and physicists incarcerated in gulags in Stalin's Soviet Union. A symposium of the International Committee of the Red Cross drew attention to similar efforts being made in the US in the 1970s.[8] Much of this research has been carried out in US universities under contracts from the somewhat sinisterly acronymed DARPA – the Defense Advanced Research Projects Agency,[9] a Federal agency which has funded much of the US research on artificial intelligence since the 1960s.[10] Some of the evidence for this interest has been collected recently by John McMurtrey for a group calling itself Christians Against Mental Slavery, and whilst I disagree with their conclusions about the prospective power of these technologies, they have done a valuable service in compiling the data.[11]

McMurtrey draws attention to a number of patents relating to such technologies. A US patent in the name of Kiyuna and others in 1998 for a system and method 'for predicting internal condition of live body'[12] is apparently intended to be used 'to detect the internal condition of surveillance in criminal investigation'. However, making such measurements in the present context demands that the subject wear electrodes or a recording helmet, and be placed in a special apparatus, like the MEG facility discussed in Chapter 6. To be of military use the technology would have to be capable of remote application. There are patents that speak to this possibility, via transmitter-capable skin implants, and neural networks. There are even speculations about the prospect of 'thought reading' or 'synthetic telepathy' as a way for intelligence agents or special forces to communicate with one another.

Closer to hand, at least in the US, some neuroimaging methods are now admissible in court. A company called Brain Fingerprinting has patented the use of EEG for this purpose. The claim, to quote from their website, is that 'brain fingerprinting testing can determine the truth regarding a crime, terrorist activities or terrorist training by detecting information stored in the brain'. The research on which this claim is based, the website announces, was funded by the CIA and FBI, and one of the company's directors, Drew Richardson, 'Vice President of Forensic Operations' (*sic*), is an ex-FBI agent. The idea that the complex patterns of EEG waves could be used to read the mind has been around ever since the 1920s, when the Swiss physiologist Hans Berger first taped a set of recording electrodes to the scalp and recorded the passionate bursts of electrical activity that rippled through the brain. As Chapter 6 describes, the wave patterns vary in wakefulness and sleep, and during focused activity. Changes in pattern can be evoked by sensory stimuli. The 'brain fingerprinting' claim is that if an individual is confronted with an object relevant to his past, and which he would be uniquely placed to remember (such as the murder weapon to an alleged murderer) then the EEG pattern will change in a characteristic way, involving a wave form known as P300, which appears in the EEG during many memory-related tasks and other forms of focused activity. The recognition of guilt is, it would appear, built into the brain waves, just as it was once believed that the image of a murderer was burned into the retina of the murder victim. Of course, a negative reaction – a failure to respond to the image – might be interpreted as evidence of innocence, and indeed the company claims that evidence based on the technique played a part in the appeal which led to the overturning of a conviction for murder in an Iowa court in 2001.[13]

The use of the term 'fingerprinting' cleverly makes an analogy with DNA fingerprinting, a far more specific, though still sometimes controversial technique By contrast, 'brain fingerprinting' would appear to be a variant on the earlier forms of lie detector, which measure the electrical potential across the skin, the galvanic skin response (GSR). The GSR depends, amongst other factors, on the ionic content of the moisture on the surface of the skin, and this in turn may vary depending on how anxious a person is. The GSR, it is assumed, will therefore fluctuate when a person under interrogation deliberately tells a lie. Unfortunately, trained liars can control their responses, whilst innocent anxious people may not be able to do so. Can 'brain fingerprinting' do any better? Frankly, I doubt it, but in the heightened atmosphere of fear and repression that has dominated the US, and to a lesser extent European countries, ever since 11 September 2001, the search for such methods of predicting, preventing or convicting alleged terrorists has been dramatically increased. The US budget for 'biodefense' has escalated, and neuroscientists have not been slow to see an opportunity to tailor their research towards this lucrative new source of income. Under such circumstances snake oil can become a valuable commodity.

The claim for such fingerprinting is that it can provide evidence as to past guilt or innocence – but might imaging also enable one to predict future behaviour? It has for instance been suggested that it might be possible to identify violent killers – 'psychopaths' – before they have actually killed. The question has been brought into sharp focus in the UK in recent years, especially in the context of the trial and conviction of Michael Stone for the brutal murder of a mother and her daughter. At his trial in 1996 Stone was judged to be sane, and was therefore imprisoned rather than placed in a psychiatric institution, but he was also said to be psychopathic – that is, capable of impulsive and apparently unprovoked violence. The murders led to a campaign to change the law so as to make it possible to identify and confine such psychopaths *before* they had committed any crime. In this context much was made of the claims that PET scans could reveal abnormalities in the brains of those judged to be psychopathic. A strong advocate of this possibility has been a former Home Office forensic psychiatrist now working in California, Adrian Raine. In the mid 1990s a UK television series, entitled *A Mind to Crime*,[14] featured Raine standing in front of two such PET scans, and explaining that one was of a psychopathic murderer, the other of a 'normal' person. The psychopath was said to show less activity in the prefrontal cortex, and other abnormalities in the amygdala. The TV presentation was derived from a study Raine had made of forty-two men all of whom had killed strangers, but were judged sane at trial, and who, he claimed, generally showed these abnormalities.[15] To support such studies, there has even been a move back to the study of dead brains. According to one newspaper report, the brains of two of Britain's most notorious serial murderers, Fred West and Harold Shipman, have been retained for 'genetic and imaging analysis'[16] – though as you don't need brain tissue to study the genes, and it is hard to envisage how one could image a dead brain, it doesn't seem likely that anything more than prurience will result.

Could such studies really be predictive? Even if differences in the scans do in fact match differences in behaviour, these are correlations, not causes. Raine's subjects were men scanned after they had murdered and been jailed. Could this experience have resulted in the changes in the scan? Even more important, there is no information on how common the alleged abnormalities might be in a population of non-murderers – 'aggressive' businessmen, 'heroic' surgeons, highly competitive sportsmen

– or for that matter in women. Third, there is no information about how many perfectly sane killers do not show such abnormalities. Would one expect to find such differences in the brain of an American pilot dropping cluster bombs from a plane in Iraq, or an Israeli officer emptying his machine-gun into the body of a Palestinian schoolgirl – or for that matter a British Prime Minister who has sent troops into battle across three continents in five years?

The judgement over whether an act of killing is legitimate, socially sanctioned, a matter of justifiable self-defence, or murder, even outside military conflict is an issue for the justice system, not biology. Consider two relevant recent British cases. In 1992 whilst on duty in Northern Ireland, British soldiers shot and killed an Irish teenager who drove a car past a checkpoint without stopping. Two soldiers were tried for murder, found guilty and jailed. A concerted press campaign led to the case being reconsidered and the sentences quashed in 1998; the soldiers were reinstated in the army and one has been promoted to lance-corporal. By 2004 they were said to be serving in Iraq. In 1999 the reclusive farmer Tony Martin shot and killed a young burglar, whom he had disturbed at his farmhouse, whilst the young man was running away. He was found guilty of manslaughter and spent three years in jail before being released on parole in 2003 after another vociferous press campaign in his support argued that he had a right to protect his property. In the same year a BBC poll was inundated with proposals that a 'Tony Martin's law' be introduced to make such killing in defence of one's property legal. Thus a given act may be legal or illegal, moral or immoral, yet the brain processes involved in performing it would be identical. No amount of imaging of brains, either before or after their owners had killed, would have helped resolve these legal and ethical dilemmas.

But let's try another of those thought experiments. Suppose that brain imaging gets better than it is at present – not too unlikely a prospect – and further, that it is then possible to perform properly designed predictive scanning which would enable one to say that such and such an individual has an increased risk of committing a violent unprovoked murder. What are the implications? Should one, as David Blunkett, the former British Home Secretary, seemed to believe, be entitled to take pre-emptive action on the basis of detected abnormalities in a person's prefrontal cortex or amygdala structure or activity? What happens to the presumption of innocence that has been a cornerstone of the legal system for centuries? Alternatively, would a person found guilty of some violent act be entitled to plead diminished responsibility on medical grounds? What do such determinist arguments do to our concepts of responsibility and of human agency? These and other ethical, legal and social issues are brought dramatically into view by this new neurotechnology – and are only a foretaste of what is to come as we neuroscientists, neurogeneticists and neurotechnologists get fully into our stride. My own answers will come in a while, but there is more ground to cover first.

Predicting the future: neurogenetics

Neuroimaging may reveal past transgressions or even predict future tendencies in a child or adult, but what if one could make the predictions much earlier, in the newborn or even before birth? These are the hopes now invested in genetics. The claim that behaviours, or at least behavioural tendencies, are inherited is of course ancient. The prospect of identifying families whose offspring might be morally or intellectually degenerate and who should therefore be discouraged from breeding by law or physical intervention was the goal of the eugenics movements of the late nineteenth and early twentieth centuries. The segregation of 'unfit' men and women to prevent their breeding in Britain, the immigration restriction and sterilisation programmes of the 1920s and 30s in the US, the compulsory sterilisation programmes in Sweden, which persisted into the last decades of the twentieth century, and above all the elimination of 'lives not worth living' in Nazi Germany all reflected these eugenic concerns. These histories have been well documented, as has the troublingly incestuous inter-relationship between eugenics and the development of genetics as a science[17] and it is not my intention to retread this ground here.

However, this history has had a profound influence on both the goals and methods of behavioural genetics, notably in its obsessive concerns with finding quantitative scales on which to measure behaviour or performance, such as IQ, and in endeavouring to separate the genetic and environmental contributions to such measures by heritability estimates. Such estimates provide bizarre reading, as it turns out that, if they are to be believed, a wide swathe of even our most seemingly trivial or personal beliefs and actions are profoundly shaped by our genes. Heritability estimates above 35 per cent have been obtained in the US for such diverse attributes as sexual orientation, alcoholism, drug addiction, 'compulsive shopping', religiosity, tendency to midlife divorce, political affiliation and attitudes to the death penalty, Sabbath observance, working mothers, military drill, royalty, conventional clothes, disarmament, censorship, 'white lies' and jazz. Even nudist camps and women judges come in at about 25 per cent, so it is perhaps a matter of surprise that there is virtually no heritability to attitudes to pyjama parties, straitjackets and coeducation.[18] But when one learns that men show a significant heritability for enjoying having their backs scratched in the bath, whereas women do not,[19] the dubiety of the measure becomes more obvious.

A somewhat more sophisticated approach to identifying genes relevant to behaviour begins with family studies in an effort to identify genes or gene markers which might be predisposing or risk factors associated with or predictive of specific behaviours. Such approaches are also fraught with problems. I've already mentioned the endless and so far fruitless hunt for genes that might predispose to schizophrenia, and the attempts to identify genes relevant to ADHD diagnoses, but even these are based on rather simplistic partitioning out of a 'behaviour' between additive contributions from 'genes' and from 'environment'. At best one gets a sort of crude version of the old claim by some social scientists that biology ends at birth, after which culture takes over. An example is provided by the child psychologist Jerome Kagan, who argues that gene differences are responsible for temperament – the variations in emotional reactivity, such as timidity and fearlessness, apparent even in very young infants.[20] Culture – environment – may superimpose character and personality on such underlying genetic givens, claims Kagan, but temperament is a gift given to the newborn child by the genetic fairy godmother. It is in the last analysis all down to different forms of the genes coding for dopamine and serotonin receptors.

Gay genes

As with brain imaging, it is primarily in the context of socially transgressive attitudes and activities that the issues of genetic predestination are posed most sharply and have been historically of most interest to behaviour geneticists. One of the most potent has been the question of sexual orientation. For decades there had been claims, based on twin and family studies, that there was a heritable component to male homosexual behaviour. In 1993 Dean Hamer, working at the US National Cancer Institute, published a paper in *Science* reporting data from a small number of families in which there were at least two declaredly homosexual brothers. He claimed to have identified a locus – a gene marker – on the tip of the X chromosome (Xq28) which was common to the gay men but not to their straight siblings.[21] The publication was accompanied, as has become common under such circumstances, with a press release which was rapidly transmuted into the announcement that Hamer had discovered 'the gay gene'. The press headlines were appropriately sensational:[22]

'It's in the genes – how homosexuals are born to be different'
  (*Daily Mirror*, 17 July)
'Genes that may chart course of sex life' (*Daily Mail*, 17 July)
'Proof of a poof' (*Sunday Sport*, 18 July)
'Mums pass gay gene to sons say doctors' (*Sun*, 17 July)
'Abortion hope after "gay genes" finding' (*Daily Mail*, 16 July)

Hamer became an instant celebrity, appearing before the press to insist that, if he were able to go further and identify not just a marker but 'the gene' itself, he would patent it to prevent its exploitation for prenatal diagnosis and hence potentially for abortion of foetuses predestined to become gay. His findings were welcomed in the US by gay organisations which argued that they proved that gayness was a natural condition, not a disease, and that gays should not therefore be criminalised. T-shirts appeared carrying the slogan 'Xq28 – thanks, Mom'. (The marker is on the X chromosome and therefore inherited maternally.) Critics wondered why there was anything to be gained from such studies, and why no one, it seemed, was interested in investigating the genetics of homophobia. Perhaps taken aback by the reception to his work, Hamer went on to argue that research on the biological underpinnings of gayness should only be undertaken by researchers who were themselves gay.

Several years down the line, both the enthusiasm and the concern have somewhat receded. As is often the case for claims to have identified genes associated with specific behaviours, Hamer's results have not been replicated in other family studies.[23] But there are more fundamental problems, as with many such attempts to treat human behaviours or mental attributes as if they were *things*, measurable phenotypes. Thus, to argue that there is a genetics of homosexuality one must begin by taking it for granted that sexual orientation is a fixed category, stable and innate. Yet historians of sexuality point out that the very concept of a straightforward bimodal (hetero/homo) distribution of human sexual orientation, fixed at conception or birth, is modern. In some cultures homosexual behaviour is a normal and accepted developmental phase. For sure the meaning and expression of both male and female homosexual behaviour has differed at different times in human history, from Plato's Athens to Oscar Wilde's London and Dean Hamer's Washington. The steps by which such same sex activity becomes objectified and then given a genetic locus exemplifies the reifying tendency of much neuroscience, and in particular neurogenetics.

Criminal genes

Gayness may still be regarded as undesirable or an aberration in some quarters – witness the agonising knots that the Anglican church has got itself into over the question of the conditions under which, and the orifices into which, their clergy are permitted to insert their penises. However, it is no longer criminalised by the courts or classified under DSM as a disease. Of greater concern for those interested in the predictive powers of genetic knowledge is the possibility of genetic predispositions to criminal and 'anti-social' behaviour. The state's interest in identifying such individuals is obvious and was reflected in the discussions about the uses of amygdalectomies discussed in Chapter 9. In the US, a specific Federal Violence Initiative aimed at identifying inner city children 'at risk' of becoming criminally violent in later life was proposed in the early 1990s by the then Director of the US National Institute of Mental Health, Frederick Goodwin, but ran into intense criticism as being implicitly racist because of its repeated coded references to 'high impact inner city youth'.[24] Plans to hold a meeting to discuss Goodwin's proposals were several times abandoned and shortly afterwards he left the directorship. However, parts of the programme have been implemented in Chicago and other big US cities, and eventually, in 1995, a modified version of the meeting was convened in London under the auspices of the CIBA Foundation.*25

The preoccupation with criminal genes goes back to the early days of eugenics, with its dubious records of family histories and twin studies. Much of the data in support of some genetic involvement comes from the comprehensive twin and adoption registers maintained in Sweden and Denmark. Analysing this data led to the conclusion that there is a statistically significant chance that if the genetic father of an adopted child has a criminal record, then the child will also show criminal or anti-social behaviour. However, it appears that this correlation only applies to property crime, and not to crimes of violence.26 Not surprisingly, this pattern seems to have puzzled the researchers. Could one envisage genes predisposing a person to be a burglar but not a robber, a pickpocket but not a mugger?

Relabelling the genetic predisposition as one to 'anti-social behaviour', rather than criminality per se, as in the title of the CIBA symposium, may seem less provocative, but it neither resolves the paradox of the Danish and Swedish studies nor does it satisfy the wish to prove a genetic association with violence. This link was provided by a report in 1993 by Han Brunner and his colleagues describing a Dutch family some of whose menfolk were reported as being abnormally violent. In particular, eight men 'living in different parts of the country at different times across three generations' showed 'an abnormal behavioural phenotype'. The types of behaviour reported included 'aggressive outbursts, arson, attempted rape and exhibitionism'.27 These individuals also all carried a mutation in the gene coding for the enzyme monoamine oxidase A (MAOA) which is involved in dopamine and serotonin neurotransmission. Lumping all of these together as manifestations of an underlying reified behaviour, 'aggression', seems dubious, and Brunner was very careful to distance himself from some of the stronger interpretations of his result. After all, the paper referred to 'abnormal' not

* The CIBA (now Novartis) Foundation was established by the drug company of the same name – the company, as mentioned above, that markets Ritalin. However, the Foundation is independent of its parent company.

'aggressive' behaviour. None the less, MAOA became widely reported as the 'violence gene'. It was followed up by studies in which the MAOA gene was deleted in mice. The deletion was lethal on homozygotes (those with both copies of the gene deleted), but with only one copy deleted (heterozygotes) the mice survived for a few weeks. The authors describe them as showing 'trembling . . . fearfulness . . . frantic running and falling over . . . [disturbed] sleep . . . propensity to bite the experimenter . . .' Of all these features, the authors chose to highlight 'aggression' in their paper's title, concluding that their results support 'the idea that the particularly aggressive behavior of the few known human males lacking MAOA . . . is a more direct consequence of MAO deficiency'.[28] Thus are myths born and consolidated.

Brunner's original work has not been replicated, but in 2002 Avshalom Caspi and his colleagues came to a more subtle conclusion, studying a thousand adults, a New Zealand cohort aged twenty-six at the time of the study, whose members had been assessed routinely from age three onwards. A proportion of these adults had been maltreated in various ways when children; as adults, a proportion were violent or anti-social, based on actual criminal convictions or personality assessments. The research found that men who had been maltreated when children were more likely to be violent or anti-social when older, an unsurprising result. The interesting feature of the study though was to relate their behaviour to MAOA activity. There are a number of known variants of the MAOA gene which affect the enzyme's activity. Childhood maltreated individuals with the variant producing high levels of MAOA were less likely to be violent or anti-social as adults than those with the gene variant producing low levels of activity. As the authors conclude, this is evidence that genotypes can moderate children's sensitivity to environmental insults.[29]

Working with the same cohort, the researchers went on to suggest that different forms of a gene associated with serotonin neurotransmission moderated the relationship between life stress and depression.[30] Should we be surprised? Surely not. As with Kagan's findings, which I discussed above, developmental systems theory would predict such interactions, but within our theoretical framework the concept of a gene in isolation from its environment, and able therefore to 'interact' with it as if the two could be arbitrarily partitioned, is simply the wrong way to think about the developmental process through which such forms of thought and action emerge. But even for behaviour geneticists, the Caspi group's findings direct attention away from simplistic assumptions about genetic predictability towards a richer recognition of the subtleties of the developmental process.[31] Behaviour genetics – even when it rids itself of its more naïve commitment to impoverished concepts of behaviour, to reductionist parcelling out genes and environment, to unreal assumptions about twins and the inappropriate use of heritability estimates – has a long way to go before it becomes a predictive science.

Changing the future: psychogenetic engineering?

Leaving mere curiosity aside, the rationale for genetic studies of behaviour must be to attempt to find ways of modifying or eliminating such behaviour if undesirable, or enhancing it if desirable. The popular and bioethical literature is replete with speculation about the prospects of such genetic manipulation. It is of course already possible to identify genes predisposing to neurological diseases such as Huntington's, and potential genetic risk factors for Alzheimer's disease, and to offer pre-natal diagnostic testing, but claims for genes associated with risk of depression or schizophrenia are still a matter of contention, even amongst those who do not doubt the specificity and accuracy of the diagnoses, and by this stage my extreme scepticism about the evidence or relevance of genes directly associated with criminality or sexual orientation should be clear. However, set my scepticism aside for a moment and take the genetic claims at their most optimistic, and consider their likely implications.

There are potentially two forms of genetic manipulation. One, germ line therapy, would involve taking the affected egg or sperm and replacing the affected allele with a 'corrected' version. This is the procedure used routinely, and with varying degrees of predictability as to the phenotypic outcome, in the 'construction' of genetically manipulated mice in the laboratory. This is what is popularly understood as implied by the term genetic engineering. The alternative is what is known as somatic gene therapy – trying to manipulate the genes in an already-developing or even adult individual by inserting new or modified DNA and hoping that it becomes incorporated into the genome of the adult cells.

Germ line therapy, which would add or eliminate genes not merely in the treated individual but in his or her offspring, is currently illegal in most researching countries. Even were it not, it is dangerous and its results unpredictable; in mice the failure rate is high, and the effects of inserting or deleting individual genes unpredictable for all the reasons I continue to dwell on. Its advocates have none the less argued that its use might be considered for dreadful disorders such as Huntington's Disease, a monogenic autosomal dominant condition. That means that only one copy of the gene, in either sex, is sufficient to predict that the person carrying it will suffer progressive and irreversible mental degeneration from some time in their middle years. Pre-natal diagnosis has been possible for several years. However, it seems that only a fraction of those people at risk of Huntington's actually opt to take the genetic test, although many may try to avoid the dilemma by choosing not to have children. There are many reasons why people might not wish to take the test. The most straightforward is that they may simply not wish to know. Genetic knowledge is special as it gives information not only about the person being tested but about the risk to those to whom they are genetically related. Learning that one carries the gene tells one therefore about one's parents and also the risk to one's children, and this is a complex responsibility to bear. Furthermore, there is no neuroprotective treatment currently available for someone carrying the gene, and the only currently preventative option would be abortion of any foetus testing positive. Nor, I would guess, is there likely to be such a treatment in the immediately foreseeable future, despite intense research on the molecular biology of the condition. So, even were it legal and available, I find it hard to imagine that any prospective parent would opt to take the genetic engineering route – and this for a condition whose dire effects are predictable and from which there is no obvious prospect of escape. So, for other late onset diseases, where there is no one single gene involved but many, each with a small risk associated, as in Alzheimer's, it is hard to imagine that germ line therapy is a serious option. Still less is this likely to be the case for schizophrenia or depression.

As for somatic gene therapy, there have been a variety of trials over the past decades, notably for genetic disorders such as cystic fibrosis and severe combined immunodeficiency disease. The results have generally not been encouraging,[32] even with these relatively well-understood monogenic conditions. When it comes to behaviour as opposed to disease risk, the stakes are even higher. If Hamer's results had been replicated, and Xq28 did turn out to be a marker for a gene associated with a tendency amongst men to homosexual behaviour in adulthood, could one really envisage potential parents going through egg selection to avoid it, still less to put a foetus through somatic cell therapy to obviate its effects, known and unknown?

A distinction is sometimes made between gene therapy to eliminate undesirable characteristics and treatment to enhance desirable ones. The line between the two is at best very thin – perhaps no more than a matter of semantics, as bioethicists who rush in to defend the prospect of enhancement are almost indecently quick to insist.[33] But the prospect of such genetic enhancement remains at the extreme periphery of science fiction. To take a potential example: it has been suggested on the basis of twin studies that a gene associated with a receptor for a particular growth factor may also be associated with a small percentage increase in IQ score.[34] Could this be the basis of genetic enhancement of intelligence? Or how about inserting the gene for the glutamate receptors shown to be associated with enhanced memory, based on Tsien's mouse work? Or CREB, based on Tully's prospectus? Suppose, first, that IQ is a useful measure of some desirable feature, and that this claim is vindicated, and that the techniques for inserting such a gene either somatically or in the germ line were available and legal – or at least purchasable in some offshore rogue laboratory, rather as the current spate of cowboy cloners claim to be operating. Grant too that all the problems of safety and of failure rates were overcome, and that the gene could be predictably inserted in the right site in an acceptably high percentage of trials. Even if the procedure were successful, would the resultant child show the desired enhancement? The first problem is that the effects of inserting such genes cannot be predictive. At best they could be considered as the equivalent of risk factors. And, like most other genes, their effects are pleiotropic, affecting many physiological processes in a way that depends on the background of the rest of the genome within which they are expressed. The receptor mechanisms that such genes are associated with involve pretty basic aspects of cellular molecular biology, and the consequences of modifying them are bound to be widespread.

These are pretty large hurdles to overcome. However, I have gloomily to conclude that – bearing in mind the current experience of the enthusiasm amongst the wealthy for so many other putatively enhancing treatments such as cosmetic surgery – if they were, there would be a potential market for such manipulation. Indeed, there is some past experience to go on. Back in 1980, a 'genetic repository' was created in California, and Nobel prizewinners and other distinguished figures were encouraged to 'donate' their sperm, which was then made available to women wishing to become the vessels through which these great men could propagate their genes and, putatively, their intelligence (to say nothing of any other tendencies: to obsessiveness, baldness, shortsightedness or the vanity to believe in the high desirability of their own sperm). The eugenicist journal *Mankind Quarterly* reported euphorically in 1993 that there had already been nearly two hundred offspring thus conceived.[35] However, whether they have inherited any of their fathers' more desirable characteristics, or whether these have been diluted by the regrettable necessity, in the absence of cloning, of combining the precious paternal DNA with 50 per cent of that from the mere mother, remains to be seen.

If snake oil is well marketed, there are always likely to be purchasers. The question is whether their money is well spent. High IQ isn't a very good indicator of social or professional success (as a perusal of the membership lists of MENSA, the club for high IQ individuals, will testify). But a child's IQ does correlate with parental income, and the best way of 'enhancing' a child's performance remains 'environmental' rather than genetic – choose a good school and devote time and energy to the educative process in its broadest sense. Ethics apart, there are no cheap genetic fixes, nor are there likely to be. Buying privilege in the old-fashioned way is likely to remain the best option for those who share Dean and Morgenthaler's view that life is a race in which the task is to get ahead of as many others as possible.

Towards Soma for all

The twenty-first-century world is one in which the chemical means for modifying perception, emotion and cognition are becoming ever more varied and ever more frequently used. It is worth pausing for a moment and asking why this should be so? Mind-changing can be a source of individual pleasure and/or a social lubricant. I would hate to live in a world devoid of wine or whisky, and others will surely feel similarly about their own favourite consciousness modulator. Such drugs can also be an essential way of easing the pain of living. The sheer scale of psychotropic use, the level of mental anguish reflected in the WHO's estimates of the world-wide incidence of depression, speaks of a situation in which our minds are seriously out of joint with the environment in which they are embedded. Also, not unimportantly, there is an increasing tendency within our highly individualised culture and society to medicalise social distress and to label unhappiness as a disease state, to be rectified by medical intervention, whether by drugs or by other forms of therapy. The US Constitution speaks of every man having the right to life, liberty 'and the pursuit of happiness'. Ignore the issues of race (it didn't mean *every* man) and gender (it did mean *man*), as Dick Lewontin pointed out many years ago.[36] It is not *happiness* the US founding fathers believed men were entitled to, merely its *pursuit*. Today, certainly in the US and to an increasing extent in Europe, we assume our right to happiness – and not happiness deferred, either, but happiness NOW.

'Happiness Now' could be a political slogan, alongside the graffito scrawled on an Oxford college wall, 'Do not adjust your mind, there is a fault in society', that many years ago, when some of this debate was in its infancy, Hilary Rose and I took as the title for a paper.[37] But mind adjustment seems easier than social adjustment; one only has to remember that mordant comment on the condition of working people in the English cities during the early industrial revolution – 'Penny drunk, tuppence dead drunk; cheapest way out of Manchester'. Aldous Huxley recognised this in his *Brave New World*, where tranquillity is ensured by the liberal availability of Soma.

In today's sophisticated neuroscientific context, Huxley's Soma, an off-the-peg universal provider of Happiness Now, is giving way to a vastly varied menu of potential psychochemicals, tailored to each person's individual preference or even genotype. Over the past century the traditional routes to mind- and mood-changing, such as alcohol, nicotine, cannabis, peyote, opium, magic mushrooms and mescaline (this latter Huxley's own choice in the years following *Brave New World*), began to be supplemented by synthetic products. To the official pharmacopoeia that I discussed in Chapter 8 must be added such substances as LSD, Ecstasy, heroin and crack cocaine, to say nothing of the glue solvents that are the last resort of the financially and pharmacologically impoverished. The legal and social status of these many agents is hopelessly confused. Some are illegal, and selling or using them is subject to varying degrees of penalty. Some are legal but obtainable only by medical prescription, while others can be purchased over the counter. Some we may be urged to take by our doctors or psychiatrists, despite the dangers; others we are strongly advised not to consume for fear of addiction or worse. Which category any agent falls into differs from country to country in Europe, and from state to state in the US. It also changes over time. Opium was once readily and legally available in Britain, now it is not. The rules concerning the purchasing and using of alcohol, nicotine and cannabis are in a state of flux. Meanwhile, many more products with inflated claims but of dubious efficacy are available via the legal penumbra of the Internet.

Bespoke prescribing

What the next decades will surely have on offer is an even greater and possibly more sophisticated range of psychochemicals. For some, this will offer the prospects of an extension to the 'turn on, tune in, drop out' project initiated by Timothy Leary and others in the heady 1960s, with the discovery of the joys of LSD, a mind-trip away from the problems and complexities of a work- (or unemployment-) stressed daily life. Increasing knowledge of the mechanics of neurotransmission may indeed enable the synthesis of rather specific psychotropics, enabling one to choose from a range of drug-induced sensations as subtly varied as those offered by a paintmaker's colour chart. Meanwhile, within the approved clinical framework, the central players are the increasingly giant pharmaceutical companies and their smaller associates, the multitudes of start-up biotech companies. For big pharma the need to discover and market novel drugs is becoming imperative as patents run out on existing products, but the costs of bringing a new agent to market, via the safety, efficacy and reliability hoops that the current legislation insists on, is ever rising. The drug companies themselves estimate the figure at around $500 million, although their critics claim this is much exaggerated. As their search – and advertising – intensifies, the soundbites become ever more explicit. In her recent book, *Brave New Brain*,[38] Nancy Andreasen, one of America's leading neuropsychiatrists and a member of the task force involved in constructing the latest versions of the DSM, proposes the discovery of 'a penicillin for mental illness' as the goal of twenty-first-century medicine. Another psychiatrist, Samuel Barondes, entitles his latest book *Better than Prozac*.[39]

The titles may be optimistic, but to some degree they are whistling in the dark. Their determined optimism springs as much from a recognition of the limits of modern psychiatry and its existing pharmacopoeia as from a clear sense that something better is on its way. Both the problem and the hope lie in the realisation that for clinically recognised conditions such as depression, anxiety or schizophrenia, to say nothing of Alzheimer's disease, the available drugs only work for a relatively small proportion of patients, and often the improvement over placebo is small. Many drugs only work for a short period, after which the medication needs increasing or changing, and some people show severe adverse reactions. When the hype dies down after every latest wonder drug launch, as it has by now with the SSRIs, there is a gloomy recognition that the newer generation of drugs is not much of an improvement over the older ones, and for at least a proportion of patients may actually be worse – a point well made by Barondes. It is this that underlies the seeming pessimism of Allan Roses' statement that I quoted in Chapter 9, although this hasn't halted the dramatic growth in the market for such drugs, which has risen in the US from $6 billion annually in 1995 to $23 billion in 2002.

Roses, Andreasen, Barondes, and many of those working in the pharmaceutical industry, however, see a saviour in sight, a saviour called pharmacogenetics. The pharmacogenetic argument is straightforward enough. Depression, anxiety, schizophrenia, may not be unitary categories. Therefore there may be a variety of different 'underlying' biochemistries associated with them, each the product of a different gene or set of genes. If one could identify for each person the specific set of genes 'predisposing' them to the condition, or perhaps predictive of an adverse reaction to a particular drug, then one could in principle produce a tailor-made prescription, one specific to that person's circumstances. This is the logical extrapolation of the *ex juvantibus* principle. To take an example not from psychotropics but from another well-known clinical condition, hypertension, there are at least three very different types of drugs available for blood pressure reduction: beta-blockers, calcium channel blockers and ACE inhibitors. Some people have adverse reactions to one type (for example, ACE inhibitors make me cough terribly) but not another (I'm OK with the calcium channel blockers).

At present there is no way of telling in advance which drug to prescribe, and patients may have to work through the series before finding the one that best fits, as I had to do myself with the anti-hypertensives. This is both slow, expensive and potentially hazardous. If a simple genetic screen were available, and thus there were clear markers as to which of the drugs is likely to be most beneficial and which would result in adverse reactions, the benefits could be substantial. Further, suppose that there are six or so different genes or gene combinations, the presence of each of which might be associated with the diagnosis of schizophrenia. Each set of genes is also associated with the production of a particular set of proteins. 'Decoding' these proteins and how they interact in the cell will provide a clue as to how one might design entirely novel drugs to modulate the functioning of the proteins. This at any rate is the hope. Note that this isn't 'genetic engineering' but using information derived from genetic studies to help understand variations in cellular metabolism.

That such information can be used in order to help rational drug design is clear. It has helped, for instance, in our own work on possible treatment for Alzheimer's disease, which I mentioned in Chapter 7. However, even though for Roses, Andreasen and Barondes, and especially for the pharmaceutical companies, pharmacogenetics is seen as the 'next big thing', I remain rather sceptical about its potential in the context of psychiatric disorders, for several reasons. Above all, despite the certainties of the psychiatric profession, depression and schizophrenia are problematic diagnoses, much dependent on medicalising assumptions and the fashions of the times, so it is not probable that there will be a clearly recognisable DNA profile associated with the diagnosis. But even if there were, the pharmacogenetic assumption that one will be able to read off a person's most beneficial drug combination from a gene scan is itself naïve, for all the reasons I have discussed in earlier chapters about the relationships between gene and phenotype. If there are only one or two relevant 'susceptibility genes', as they are called, then it may be possible; but if, as is more likely to be the case for most complex experiences of mental distress, there are many genes each of small effect, in combinatorial relationships with the environment during development, then a simple scan is going to lack predictive power. It would be as useful as trying to read the future in the tea-leaves at the bottom of a cup. One would be better advised to rely on clinical judgement, fallible though that clearly is.

All this makes it rather doubtful that psychopharmacogenetics will provide the hoped-for solution to the limitations of current generations of medically-used psychotropics, but there is also another, more general problem with pharmacogenetics, whether for problems of brain or body. The great merit of Huxley's Soma was that it was indeed a universal drug that could be mass-produced. The more the new genetics offers the possibility of moving away from universalism into individually tailored prescriptions, the greater the problems for the drug companies. Granted the resources required to bring a new drug to market, the companies have to be able to predict large sales if they are to make a profit, but with the advent of pharmacogenetics, instead of being able to rely on the mass sales of a limited number of drugs, they will be in the position of having to produce small quantities of a much greater range of substances. Economies of scale will disappear and the costs of the bespoke drugs may thus put them out of reach. However, the development of genetic screens which might provide warning of potential adverse reactions in a person about to be prescribed the drug could be beneficial, both in protecting the person concerned, and, from the commercial point of view of the drug companies, protecting the companies against litigation.

When Soma fails . . . the arrival of the calmatives

The goal of prediction is to control, to modify or prevent adverse outcomes, and to increase the prospect of positive ones. Childrearing, education, socialisation, the criminal courts, are all, despite their obvious differences, traditional means to these ends. Realistically, even the best devised of neurotechnologies are not likely to supplant them – but they are likely to enhance their power and efficacy. Such aspirations can be entirely well meaning and potentially beneficent, as when a company formed by prominent US neuroscientists offers the prospect of 'better reading through brain research' for children suffering from dyslexia,[40] or when it is suggested that drugs which erase recent memories might be beneficial in treating people suffering from yet another of those new diagnostic categories, Post-Traumatic Stress Disorder,[41] or, at the opposite extreme, they can be aimed at suppressing unwanted behaviour, from disruption in class to mayhem on the streets.

Much of any future increase in the power to control and manipulate is going to come from the pharmaceutical industry, and the scale and potential of what might be on offer can be gauged by extrapolation from the drugs I've discussed already: SSRIs, Ritalin, cognitive enhancers and their ilk are going to remain the most common routes to psychocivilisation for the indefinite future. However, in a society committed either to maximising the freedom of its citizens, or at the least, sophisticatedly minimising social discontent, it will always be best to allow people to find their own most effective form of Soma, from champagne to cannabis, rather than to administer a one size fits all external dose.

Huxley's Soma was a way of ensuring that the masses were tranquil, satisfied with their lot in life. For many years, however, it has been clear that there is a flip side to this type of social control mechanism. The first steps in eliminating dissent by chemical means were crude enough. The 1914–18 war saw the use of bodily poisons such as chlorine and mustard gas; later, the Nazi concentration camps used a form of cyanide – Zyklon B – to kill their inmates *en masse*. But newer and more subtle means were already under development. Unknown to the allies aligned against them, German chemists had, even before 1939, developed the first of the nerve gases, substances which can be absorbed through the skin and which, in minute concentrations, block acetylcholine neurotransmission at the junctions between nerve and muscle, resulting in swift and agonising death. These gases, Sarin, Soman and Tabun, were not in fact then used, for reasons which remain debatable – possibly fear of retaliation. In the 1950s, however, research at Britain's chemical warfare establishment at Porton Down built on this first generation of nerve gases to produce even more toxic variants, the V agents, related to commercial insecticides. Although the V agents were stockpiled in both the UK and US, during the next few decades, they have never been used in war.[42] Sarin, however, has, killing and injuring hundreds when released into the Tokyo metro system by a fanatical Japanese sect, and, above all, killing over 5000 civilians when Saddam Hussein's forces bombed the Kurdish village of Halabja in northern Iraq in 1988.*

_____

* The chemical, manufacturing and delivery knowhow and equipment were

Poisoning your enemies has long been regarded as a breach of the international laws of war and, indeed, chemical weapons were declared illegal under a Geneva Protocol signed as far back as the 1920s. The more subtle use of neurochemicals would be to find a non-lethal way of immobilising the opposition. The first such means did not directly affect the nervous system, but the early teargases were improved, again by Porton researchers in the 1950s who produced a substance code-named CS which, dispersed as a form of spray or smoke, causes retching, sweating and crying. CS and its successors became standard police issue in many countries, and they were used extensively (and certainly not non-lethally) by the US in the Vietnam war to 'smoke out' Vietnamese fighters from caves and tunnel systems, and bring them within range of more conventional weapons.

But even CS might seem a trifle crude, and, from the 1950s on, the US Defense Department (DoD) began to experiment with more specific psychochemicals. The aim was to find a substance which could be used either militarily, to disorient enemy forces, or as a means of domestic crowd control to support or replace CS. It was a period of intense interest in mood changers and hallucinogens such as LSD, itself the subject of much covert experimentation by the CIA.[43] The DoD came up with a substance codenamed BZ,* which affects another form of acetylcholine neurotransmission, not peripherally like the V agents, but centrally, in the brain itself. BZ was claimed to have the merit of disabling both hostile crowds and military personnel by causing them to become disoriented and too relaxed to obey orders. The DoD's Chemical Warfare Division produced a propaganda film showing a platoon of soldiers, ostensibly under the influence of BZ, disregarding orders, unable to fasten their uniforms and throwing down their guns. Despite claims that BZ was used in Vietnam and formed part of the stockpiles of both the Iraqi and Yugoslav army during the spate of

---

provided by European and US companies and, as Saddam was at that time supported by the US, official protests about Halabja were distinctly muted.

* The exact chemical nature of BZ was secret, but I discovered it by accident in 1979 when I was searching for a chemical blocker of this form of acetylcholine neurotransmission for some experiments I was running with my chicks. The substance I needed (its chemical initials are QNB) was provided by a colleague who slightly shamefacedly explained to me that he had himself received it from Porton.

Balkan wars of the 1990s, I am not sure that there is any credible evidence of it being used.

Neuroscience has moved on since the 1970s, and research into potential crowd control agents has continued quietly in many countries, notably the US and the then Soviet Union. Although lethal chemical weapons were supposedly banned under the Geneva Protocol and subsequent additional conventions, alleged non-lethals, for domestic use, are still seen as a legitimate part of the state's armoury of control of its unruly citizenry. The new generation of agents were comfortingly called 'calmatives' and their development went largely unnoticed until October 2002, when a group of Chechen militants entered the Moscow NordOst theatre and took the entire audience hostage. After a period of failed negotiations, the Russian military made a botched attempt to rescue the hostages by pumping in an allegedly non-lethal calmative before entering the theatre and killing the hostage-takers. Tragically, more than 120 of the hostages were overcome by the spray and died; many others have suffered permanent brain damage. The Russian authorities were very reticent about the nature of the agent they employed – even to their own medical teams, who initially thought it was related to the anticholinergics like BZ. However, the best current guess is that the substance was fentamine, an opiate-type agent.

Whether the agents in the Western stockpiles are more subtle than this is an open question. What the Moscow theatre tragedy reveals quite clearly, however, is that in practice there is no such thing as an effective but non-lethal calmative agent. All such substances have a relatively narrow concentration window between being ineffective and being lethal. Furthermore, like any other drug, the effect on individuals can be very varied, depending on their age, state of health and indeed motivational state. Psychochemicals are notoriously varied in their effect; you only have to think of the different reactions any one of us can have to the same dose of alcohol depending on our mood, the context and the company we are in at the time. When we self-administer such substances, we usually try to regulate our intake to achieve the desired effect; when they are applied externally we have no such control. To achieve its intended purpose, a calmative must be used at a high enough concentration to incapacitate all those it is aimed at without doing lasting damage to any of them. Both toxic and calming effects are normally assessed in trial with healthy young soldier volunteers under controlled

conditions and with the possibility of escape. None of these can repli-
cate the real-life situation of a varied, confused and weakened group of
people in a confined space which the Moscow theatre represented.

That such calmatives will continue to fascinate the military and anti-
terrorist theorists, especially in the current climate of fear, is sure. The
belief that neuropharmacologists will be capable of delivering such
perfect agents belongs to the fantasy world that led the US administration to predict, prior to March 2003, that their troops would be greeted
by the Iraqis as liberators. However, it is safe to predict that promises
will continue to be made, and that law enforcement agencies and the
military will continue to invest in them.

Thought control

Chemicals are not the only route by which it may be possible to modify
thoughts, emotions and behaviour. EEG and MEG record the constant
chatter of neurons as they transit information. It follows, therefore, that
interfering with these electrical interactions will disrupt or distort the
information flow. In a crude way this is exactly what is done in ECT –
electroconvulsive shock therapy, a technique which has been around
since the middle of the last century and became familiar to many
hundreds of thousands of filmgoers through its portrayal in the Jack
Nicholson movie *One Flew Over the Cuckoo's Nest*. ECT puts a blast of
electricity across the brain, temporarily disrupting all communication,
killing some cells, erasing recent memory and generally inserting a large
clumsy spanner into the brain's machinery. Its use in the treatment of
psychiatric patients, notably those severely depressed, has been controversial since it was first employed. There is some evidence that it increases
the rate at which people recover from depression, though the long-term
follow-up studies have on the whole not been very encouraging, but for
any neuroscientist, the thought of such a massive, if brief, intervention
into the delicate workings of the brain must be a cause of concern,
however well intentioned the therapeutic goals might be.

As the complex wave forms of the EEG have become easier to interpret, and their sources within the brain's structures more certain, the
idea that it might be possible to manipulate them somewhat more subtly,
either by electrical or magnetic stimulation, has gained popularity, both

as a research tool and, potentially, for therapeutic purposes. Transcranial magnetic stimulation involves the placement of a cooled electromagnet with a figure-eight coil on the patient's scalp, and rapidly turning the magnetic flux on and off. This permits relatively localised stimulation of the surface of the brain. The effect of the magnetic stimulation varies, depending upon the location, intensity and frequency of the magnetic pulses. Proponents of its use suggest that it could be employed to treat depression, or even schizophrenia, intractable pain, obsessive-compulsive disorder, or epilepsy. DARPA has let research contracts to study its use to fight fatigue in pilots, and there are claims that it enhances cognition, at least temporarily.[44] This may sound a pretty familiar list, adaptable almost at will for any new wonder neurotechnology, and one is entitled to a little scepticism.

Perhaps more intriguing are the claims that focused magnetic pulses to specific brain regions can temporarily affect thought processes, turning them on or off at the behest of the experimenter. It is above all the military that has been interested in this prospect. Just as it might be possible to use EEG patterns to read thoughts, might it also be possible actually to insert thoughts or insert instructions to the subject? Of course, if the subject has to wear a hairnet full of electrodes, or sit in a magnetically shielded room with a bucket of liquid helium on his head, then it is hard to take seriously as a military technology – but what if it were possible to achieve the same effect at a distance? Examination of DARPA contracts and patents suggests a continuing interest in this possibility, and it has clearly long fascinated the technologists.[45] In the 1970s the US Senate was told that Soviet research had identified microwave technology enabling sounds and possibly even words which appear to be originating intracranially to be induced by signal modulation at very low average power densities.[46] McMurtrey quotes the report as claiming that it shows 'great potential for development into a system for disorientating or disrupting the behavior patterns of military or diplomatic personnel'. Assessment by the Army Mobility Equipment Research and Development Command describes microwave speech transmission with applications of 'camouflage, decoy, and deception operations'.[47] 'One decoy and deception concept . . . is to remotely create noise in the heads of personnel by exposing them to low power, pulsed microwaves . . . By proper choice of pulse characteristics, intelligible speech may be created.' By 2001 the US Air Force was claiming that 'It would also appear possible to create high-fidelity speech in the human body, raising the possibility of covert suggestion and psychological direction . . . If a pulse stream is used, it should be possible to create an internal acoustic field in the 5–15 kilohertz range, which is audible. Thus it may be possible to "talk" to selected adversaries in a fashion that would be most disturbing to them.'[48] Reading these claims and their associated patents tells one a good deal about the military mind-set, but I strongly suspect that, as with DARPA's previous fascination with artificial intelligence, we are still pretty much in the world of science fantasy here. That microwaves or pulsed magnetic fields can disorientate is clear. That the technology might be developed to enable them to do so at a distance is presumably a prospect – I would need a better grasp of physics than I possess to evaluate this. But that such a technique could be sufficiently well controlled as to insert or manipulate specific thoughts and intentions in particular individuals at a distance remains in my view a neuro-scientific impossibility. It would be better to try telepathy – and don't imagine that the military hasn't!

Transcranial magnetic stimulation remains an intriguing research tool with possible, though unproven, therapeutic potential. I remain doubtful about it as a means of abusing human rights or specifically manipulating thought, not because I naïvely trust the good intentions of those researchers working to DARPA's brief, but because on the whole I think the brain doesn't work that way, and that, sadly, for the foreseeable future, there are likely to be more effective ways of achieving even the most sinister of intentions.

Biocybernetics

There is nothing new about enthusiasm for interfacing humans directly with machines, artificial intelligence or even artificial life. Hero's robots in ancient Greece, the medieval Cabbalists' Golems, Mary Shelley's Frankenstein's humanoid creation – all speak to a continued fascination with the prospect. Twentieth-century science fiction writers followed in these well-trodden footsteps, from HG Wells to William Gibson's *Neuromancer*, becoming steadily more sophisticated as the bio-, info- and nano-sciences seem to have followed where fantasy has led. Two themes run through the literature, both fiction and faction. In the one, humans are replaced entirely by intelligent artefacts, sometimes ascribed not merely intelligence but life itself. In the other, machine and human merge, with implanted prostheses supplementing or replacing the human senses, creating cyborgs or bionic people, and/or humans directly controlling the machines by thought alone without the need for physical interaction.

Today's approach to 'artificial intelligence' actually dates back to the mathematician Alan Turing and the Bletchley Park cryptographers of the Second World War. Shortly before his suicide in 1950, and at the dawn of the computer age, Turing proposed his famous test for computer consciousness. Suppose you were in communication, via a teletype, with a second teletype in an adjoining room. This second teletype could be controlled either by another human or a machine. How could you determine whether your fellow communicant was human or machine? Clearly the machine would have to be clever enough to imitate human fallibility rather than machine perfection for those tasks for which machines were better than humans (for example, speed and accuracy of calculating) but equally the machine would have to do as well as a human at things humans do supremely – or else find a plausible enough lie for failing to do so. This is the nub of the so-called Turing Test, and he believed that 'within fifty years' a computer could be programmed to have a strong chance of succeeding.[49] By the 1960s, psychologist Stuart Sutherland was predicting that within a few years computers would be given the vote.

To create a machine that could pass the Turing Test has become the holy grail for the generations since 1950 of those committed to the pursuit of what they call, modestly enough, artificial intelligence. Whether such a computer would be conscious, or intelligent, does not really concern me here, despite the gallons of philosophical ink that has been poured over the question in the last decades. For the AI enthusiasts, the problem was to decide on the best approach. From the beginning, there were two contrasting lines of thought. One either attempted to model neurons as realistically as possible, or to specify output functions – that is, the behaviours that the machine was supposed to be able to perform. You don't necessarily need to model the visual cortex to create a device which can detect shapes, colours, motion and so forth. Modellers explored both, though most of DARPA's investment went into the first, whilst neuroscientists, interested in whether such models could tell one anything about how real brains worked, favoured the modelled 'real neurons'.

This is not the point at which to explore the tangled subsequent history of AI. It is enough to say that half a century later Bill Gates has become one of the world's richest men, computers have become many orders of magnitude more sophisticated, and 'intelligent devices' are installed into even the humblest piece of domestic equipment, – and yet the idea of computer consciousness seems as chimerical as ever,[50] even when enthusiasm is undimmed.[51] In a sense it doesn't matter; the debate has become irrelevant. We have become irretrievably locked into a world in which we are at the mercy of our machines. Witness the billions of pounds and dollars spent to avoid an information melt-down as the year 2000 approached. Even the humblest act of creation now demands intelligent technology. I delight in my Mac laptop, but if it malfunctions, I cease to be able to write. The fantasy scenarios offer even more delegation to the machine, as when a couple of years ago British Telecom engineers speculated about producing a system whereby a person could 'download their memory into a chip', as if computer and brain memory were synonymous, differing only in that one is constructed out of silicon chemistry and the other out of carbon. Would our brains then be empty, merely erased discs?

The human-machine interface need not terminate at the surface of the body. Prosthetic implants, once a fantasy of television series featuring 'the bionic man' are becoming feasible. Their initial use is and will be to substitute for brain and nervous damage. Arrays of light sensors can be connected by way of implants into the visual cortex, permitting some direct visual perception bypassing damaged eyes or optic nerves. Similar arrays could help counter deafness. On the motor side, it may be possible to compensate for spinal and peripheral nerve damage, making it feasible to move a limb by 'willing'. The indefatigable exponent of such techniques is the professor of cybernetics Kevin Warwick, who has gone so far as to have implants into his arms which apparently permit him to open doors merely by waving at them.[52]

The next step is clearly enhancement. DARPA has been interested in developing helmets which would enable troops in battle to receive and integrate distant signals, to 'call up' required information, and even to transmit signals by thought alone – a sort of computer-aided telepathy. To match them will come the technology for hands-off control of instruments, or psychokinesis. Such devices to augment the senses, or to manipulate the environment, already existing in prototype form, are likely to follow a well-trodden path in moving from military to civilian use over the coming decades.

One consequence of such developments, already apparent in embryo, is a steady merging of the real and the virtual. Children now spend so many hours playing with computers or watching video images that their direct engagement with the 'real' world must be diminished. Indeed it becomes doubtful what constitutes reality any more, as when the French philosopher Jean Baudrillard, who has spoken of 'the murder of the real',[53] notoriously argued that the 1991 Gulf war did not occur, except as a sort of endless series of computer arcade images. Such a merging, such a change in the balance of sensory inputs and motor activities will inevitably also impact directly and measurably on brain structures. The plasticity of the brain, especially during development, means that intensely exercised functions occupy more brain space, relatively neglected ones being accordingly reduced. Just as taxi driving in London increases the size of the posterior hippocampus, so the fingering of a computer keyboard increases the brain's representation of the relevant digits – and, in the longer term, such technological change in turn drives evolutionary processes, by altering selection pressures and changing the definitions of favourable versus unfavourable genotypes.

What we are witnessing is the coming together of a number of disparate technologies derived from genetics, neuroscience and the information sciences which, separately but ever more synergistically, have the potential to alter profoundly not just the shape of our day-to-day lives and of the societies in which we are embedded, but the future direction of humanity itself. It is this that has led to speculations about redesigning humans,[54] the updating of HG Wells's time machine, speculation about the creation of a new 'gene rich' species[55], or the imminence of a 'posthuman future'.[56] It is in response to these prospects that the term 'neuroethics' has been invented, and it is to neuroethics itself that finally I turn.

CHAPTER 12

Ethics in a Neurocentric World

THE CONCERNS THAT HAVE OCCUPIED THE PRECEDING ELEVEN CHAPTERS have nagged away at me for the best part of the past decade, as the neuroscience whose excitement has shaped my researching life seemed to be segueing seamlessly into neurotechnology. The claims of some of my colleagues have become ever more ambitious and comprehensive. Human agency is reduced to an alphabet soup of As, Cs, Gs and Ts in sequences patterned by the selective force of evolution, whilst consciousness becomes some sort of dimmer switch controlling the flickering lights of neuronal activity. Humans are simply somewhat more complex thermostats, fabricated out of carbon chemistry. In the meantime, I have been becoming uncomfortably aware that the issues raised in the shift from neuroscience to neurotechnology have close and obvious precedents.

By the time that the National Institutes of Health in the US had entitled the 1990s 'The Decade of the Brain', advances in genetics and the new reproductive technologies were already beginning to generate ethical concerns. Back in the 1970s, when the prospects of genetic manipulation of micro-organisms – though not yet of mammals – emerged, concerned geneticists called a conference at Asilomar, in California, to consider the implications of the new technology and draw up guidelines as to its use. They called a temporary moratorium on research whilst the potential hazards were investigated. However, the moratorium didn't last – before long the prospects of fame and riches to be made from the new technologies became too tempting, the hazards seemed to have been exaggerated, and the modern age of the commercialisation of biology began. By the 1990s the biotech boom was well under way, genes were being patented, the ownership of genetic information was under debate, and speculations about the prospects of genetically engineering humans were rife. As social scientists, philosophers and theologians began to consider the implications of these new developments, a new profession was born, that of bioethicist, charged, like any preceding priesthood, with condoning, and if possible blessing, the enterprise on which the new biology was embarked. In all the major researching nations, bioethics committees, either institutionalised by the government, as in the US or France, or by charitable foundations, as in the UK, began to be established, with the task of advising on what might or might not be acceptable. Thus somatic genetic therapy was deemed acceptable, germ line therapy unacceptable. Therapeutic cloning is OK, reproductive cloning is not. Human stem cell research is permissible in Britain, but not (with certain exceptions) under Federal regulations in the US.

This experience helps explain why it was that in 1990, when the Human Genome Project (HGP) was launched, US and international funding agencies – unlike the private funders against whom they were competing – took the unprecedented step of reserving some 3 per cent of the HGP's budget for research on what they defined as the ethical, legal and social implications of the project. The acronym that this has generated, ELSI, has entered the bioethicists' dictionary (although Europeans prefer to substitute A, meaning Aspects, for the I of Implications). Admittedly, some of those responsible for ring-fencing this funding had mixed motives in so doing. Central to the entire project was James Watson, co-discoverer of the structure of DNA and for half a century the *éminence grise* of molecular biology, who, with characteristic bluntness, referred to it as a way of encapsulating the social scientists and ethicists and hence deflecting potential criticism. None the less, whatever the complex motivation that led to these programmes being set up and funded, ELSI/ELSA is here to stay.

As a neuroscientist, I watched these developments with much interest; it seemed to me that my own discipline was beginning to throw up at least as serious ethical, legal and social concerns as did the new genetics. But such was and is the pace of genetic research, and the lack of control over the work of the biotech companies, that the various national ethics councils were often at work closing the stable door after the horse – or in this case the sheep – had bolted. The successful cloning of Dolly, announced to great fanfares of publicity in 1996, took the world seemingly by surprise, prompting a flurry of urgent ethical consultations and hasty efforts to legislate. No one, it seemed, had thought the cloning of a mammal possible, although in fact the basic science on which the success was based had been done years before.

Comparable technological successes in the neurosciences are still in the future. Whereas in the case of genetics ethical considerations often seemed to occur *post hoc*, in the neurosciences it might be possible to be more proactive – to find ways of publicly debating possible scientific and technological advances before they have occurred, and engaging civil society in this debate. Over the last five years my own concerns have become more widely shared. A new term has appeared in the bioethical and philosophical literature: neuroethics. In Britain the Nuffield Council on Bioethics has held consultations on the implications of behaviour genetics. In the US, the President's Bioethics Council has discussed topics as disparate as cognitive enhancers and brain stimulation. The Dana Foundation, a US-based charity with a European branch, has sponsored discussions on neuroethics as a regular component of major neuroscience meetings.[1] The European Union has organised its own symposia.[2] Critiques of both the power and pretensions of neuroscience have appeared.[3] As I was completing this very chapter, a Europewide citizens' consultation on neuroethics, scheduled to last for the next two years, was announced.

The issues raised in such discussions range from the very broadest to the most specific. Can we, for instance, infer some universal code of ethics from an understanding of evolutionary processes? What are the boundaries between therapy and enhancement – and does it matter? How far can biotechnology aid the pursuit of happiness? Could one – should one – attempt to prevent ageing and even death? To what extent should neuroscientific evidence be relevant in considering legal responsibility for crimes of violence? Should there be controls on the use of cognitive enhancers by students sitting examinations? On what issues should governments and supranational bodies endeavour to legislate?[4]

As usual, novelists and movie-makers have got here before us. The daddy of them all is of course Aldous Huxley, but the cyborgian worlds of William Gibson's *Neuromancer* (published twenty years ago) and *Virtual Light* offer prospects that many will see as potential if unpalatable. Recent films have trodden similar paths. Thus Charlie Kauffman's *Eternal Sunshine of a Spotless Mind*, released in 2004, plays with the prospect of a neurotech company (Lacuna Inc) specialising in erasing unwanted memories via transcranial brain stimulation.

It is far from my intention to pronounce dogmatically on these quite deep questions. This isn't simply a cop-out with which to end this book; it is that I don't think that I have the right or authority to do so. They are, after all, the domain of civil society and political debate. My task as an expert with special competence in some of these areas is to lay out as clearly as I can what seem to me to be the theoretical and technical possibilities, which is what the preceding eleven chapters have attempted to do, and then to participate, as a citizen, in discussing how we should try to respond to them. What is certain is that society needs to develop methods and contexts within which such debates can be held. The processes that have been attempted so far include the establishment of statutory bodies, such as, in the UK, the Human Genetics Commission (HGC), and various forms of public consultation in the form of Technology Assessment panels, Citizens' Juries and so forth. None are yet entirely satisfactory. Engaging 'the public' is not straightforward, as of course there are multiple publics, not one single entity. Furthermore, governments are seemingly quite willing to bypass their own agencies, or set aside public opinion, when it suits. Legislation permitting the use of human stem cells in Britain bypassed the HGC. When the public consultation on GM crops produced the 'wrong' answer – an over-whelming public rejection – the government simply ignored it and autho-rised planting.

Nonetheless, the very recognition that public engagement is essen-tial, that technologies cannot simply be introduced because they are possible, and might be a source of profit for a biotech company, becomes important. In Europe we are becoming used to developing legislative frameworks within which both research and development can be regu-lated. The situation is different in the US where, although the uses of federal funds are restricted, there is much more freedom for private funders to operate without legal restriction – as the flurry of claims concerning human cloning by various maverick researchers and reli-gious sects has demonstrated. This is perhaps why Francis Fukuyama, in contrast to other American writers concerned with the directions in which genetics and neuroscience might be leading, argues in favour of the European model.

The framework within which I approach these matters is that presaged in earlier chapters. It pivots on the question of the nature of human freedom. I have tried to explain why, although I regard the debates about so-called 'free will' and 'determinism' as peculiar aberrations of the Western philosophical tradition, we as humans are radically undetermined – that is, living as we do at the interface of multiple determinisms we become free to construct our own futures, though in circumstances not of our own choosing.[5] This is what I was driving at in Chapter 6 with my account of the limited power of the imaginary cerebroscope.

We are both constrained and freed by our biosocial nature. One of these enabling constraints is evolutionary. Evolutionary psychologists, in particular, have claimed that it is possible to derive ethical principles from an understanding of evolutionary processes, although many philosophers have pointed out with asperity that one cannot derive an ought from an is.[6] I don't want to get hung up on that here; to me the point is that no immediately conceivable neurobiological or genetic tinkering is going to alter the facts that the human lifespan is around a hundred plus or minus a few years, and that human babies take around nine months from conception to birth and come into the world at a stage of development which requires many years of post-natal maturation to arrive at adulthood. These facts, along with others such as our size relative to other living organisms and the natural world, our range of sensory perception and motor capacities, our biological vulnerabilities and strengths, shape the social worlds we create just as the social worlds in turn impact on how these limits and capacities are expressed. This is the evolutionary and developmental context within which we live and which helps define our ethical values. There may be a remote cyborgian future in which other constraints and freedoms appear, but we may begin to worry about those if – and it is a disturbingly large if – humanity manages to survive the other self-inflicted perils that currently confront us. But we also live in social, cultural and technological contexts, which equally help define both our self-perceptions and ethical values. This is what the sociologist Nikolas Rose is driving at when he refers to us, in the wake of the psychopharmacological invasion of our day-to-day lives, becoming 'neurochemical selves', in which we define our states of mind and emotion in terms of medical categories and the ratios of serotonin to dopamine transmitters in our brains.

As I argued in the two preceding chapters, the neurosciences are moving us towards a world in which the prospects both for authoritarian control of our lives, and the range of 'choices' available for the relatively prosperous two-thirds of our society are increasing. On the one hand we have the prospect of profiling and prediction on the base of gene scans and brain imaging, followed by the direct electromagnetic manipulation of neural processes or the targeted administration of drugs to 'correct' and 'normalise' undesirable profiles. On the other, an increasing range of available Somas to mitigate misery and enhance performance, even to create happiness.

There's always a danger of overstating the threats of new technologies, just as their advocates can oversell them. The powers of surveillance and coercion available to an authoritarian state are enormous: ubiquitous CCTV cameras, hypersensitive listening devices and bugs, satellite surveillance; all create an environment of potential control inconceivable to George Orwell in 1948 when he created *Nineteen Eighty-four*'s Big Brother, beaming out at everyone from TV cameras in each room. The neurotechnologies will add to these powers, but the real issue is probably not so much how to curb the technologies, but how to control the state. As for thought control, in a world whose media, TV, radio and newspapers are in the hands of a few giant and ruthless global corporations, perhaps there isn't much more that transcranial brain stimulation can add.

The same applies to arguments about enhancement. In Britain today, as in most Western societies (with the possible exception of Scandinavia), enhancement is readily available by courtesy of wealth, class, gender, and race. Buying a more privileged, personalised education via private schools, and purchasing accomplishments for one's children – music teaching, sports coaching, and so forth – is sufficiently widespread as to be taken almost for granted except amongst those of us who long for a more egalitarian society. No major political party in power seems prepared to question such prerogatives of money and power. Compared with this, what difference will a few smart drugs make?

But of course, they will – if only because we seem to feel that they will. Incremental increases in the power of the state need to be monitored carefully: think of the widespread indignation at the new US practice of fingerprinting, or using iris identification techniques, on foreign passport holders entering the country, or the hostility in the UK to the introduction of identity cards carrying unique biodata. Similarly with the cognition-enhancing drugs – as with the use of steroids by athletes, their use, at least in a competitive context, is seen as a form of cheating, of bypassing the need for hard work and study. Enthusiasts are too quick to dismiss these worries, deriding them as hangovers from a less technologically sophisticated past (as when slide rules used to be banned from maths examinations). They speak of a 'yuck factor', an initial response to such new technologies that soon disappears as they become familiar, just as computers and mobile phones are now taken-for-granted additions to our range of personal powers. But we should beware of enthusiasts bearing technological gift-horses; they do need looking at rather carefully in the mouth, and it should be done long before we ride off on their backs, because they may carry us whither we do not wish to go. For, as I continue to empha-sise, the dialectical nature of our existence as biosocial beings means that our technologies help shape who we are, reconstructing our very brains; as technology shifts so do our concepts of personhood, of what it means to be human.

Which brings me, at last, to the issues of freedom and responsibility. It is in the context of the law that these concepts are most acutely tested. For a person to be convicted of a crime he or she must be shown to be responsible for their actions, that is, to be of sound mind. As Judge Stephen Sedley explains:

> When a defendant claims in an English Court that a killing was not murder but manslaughter by reason of diminished responsibility, the judge has to explain to the jury that: 'Where a person kills or is party to the killing of another, he shall not be convicted of murder if he was suffering from such abnormality of mind (whether arising from a condition of arrested or retarded development of mind or any inherent causes or induced by disease or injury) as substantially impaired his mental responsibility for his acts or omissions in doing or being a party to the killing.'[7]

As Sedley goes on to point out, the questions of 'abnormality of mind' and 'mental responsibility' raise endless tangles. They are enshrined in English law by way of the McNaghten rules under which an insanity plea can be entered, laid down by the House of Lords in 1843 and refined, as in the quote above, in 1957. From a neuroscientific point of view the definitions make little sense – as Sedley would be the first to agree. If, for instance, presence of a particular abnormal variant of the MAOA gene predisposes a person to a violent temper and aggressive acts, can the possession of this gene be used as a plea in mitigation by a person convicted of murder? If Adrian Raine were right, and brain imaging can predict psychopathy, could a person showing such a brain pattern claim in defence that he was not responsible for his actions? Can a person committing homicide whilst under the influence of a legally prescribed drug like Prozac claim that it was not he but the drug that was responsible for his behaviour? As I've pointed out, such defences have been tried in the US, and at least admitted as arguable in court, though not yet in the UK.

Such genetic or biochemical arguments seem to go to the core of our understanding of human responsibility, but they raise profound problems. If we are neurochemical beings, if all our acts and intentions are inscribed in our neuronal connections and circulating neuromodulators, how can we be free? Where does our agency lie? Once again, it is instructive to turn the argument around. Back in the 1950s it became fashionable to argue that many criminal acts were the consequences of an impoverished and abused childhood. Is there a logical difference between arguing 'It was not me, it was my genes', and 'It was not me, it was my environment'? I would suggest not. If we feel that there is, it is because we have an unspoken commitment to the view that 'biological' causes are more important, somehow *more* determining, in a sense, than 'social' ones. This is the biological determinist trap. Yet the study by Caspi and his colleagues of the ways in which gene and environment 'interact' during development in the context of any putative relationship between MAOA and 'aggressive behaviour' points to the mistake of giving primacy to a genocentric view.

Of course, it is probable both that many people with the gene variant are not defined as criminal, whilst most people who are defined as criminal do not carry the gene. At best such predictors will be weakly probabilistic, even if one sets aside the problems I raised in the previous chapter about the impossibility of extracting definitions of either 'aggression' or 'criminality' from the social context in which specific acts are performed. It follows that within a very broadly defined range of 'normality', claims of reduced responsibility on the basis of prior genetic or environmental factors will be hard to sustain. And there is a final twist to this argument. People found guilty of criminal acts may be sentenced to prison. So what should be done with persons who are found not guilty by virtue of reduced responsibility? The answer may well be psychiatric incarceration – treatment, or restraint if treatment is deemed impossible, rather than punishment. Assuming that prison is intended to be rehabilitative as well as retributive, defining the difference between the consequences of being found responsible rather than irresponsible may become a matter of semantics.[8] Despite the increasing explanatory powers of neuroscience, I suspect that many of these judgements are best left to the empirical good sense of the criminal justice system, imperfect, class-, race- and gender-bound though it may be.

But, more fundamentally, the question of how we can be free if our acts and intentions are inscribed in our neurons is a classic category error. When Francis Crick tried to shock his readers by claiming that they were 'nothing but a bunch of neurons',[9] he splendidly missed the point. 'We' are a bunch of neurons, and other cells. We are also, in part by virtue of possessing those neurons, humans with agency. It is precisely because we are biosocial organisms, because we have minds that are constituted through the evolutionary, developmental and historical interaction of our bodies and brains (the bunch of neurons) with the social and natural worlds that surround us, that we retain responsibility for our actions, that we, as humans, possess the agency to create and recreate our worlds. Our ethical understandings may be enriched by neuroscientific knowledge, but cannot be replaced, and it will be through agency, socially expressed, that we will be able, if at all, to manage the ethical, legal and social aspects of the emerging neurotechnologies.

References

Place of publication is London, unless otherwise stated.

Chapter 1

1 It seems sensible to group all these quotes together: 'Better Brains', *Scientific American*, September 2003; Greenfield, S, *Tomorrow's People: How 21st Century Technology Is Changing the Way We Think and Feel*, Allen Lane, 2003; Fukuyama, D, *Our Post-Human Future: Consequences of the Biotechnology Revolution*, Profile Books, 2002; Rose, N, 'Becoming neurochemical selves', in Stehr, N, ed., *Biotechnology between Commerce and Civil Society*, Transaction Press, New Brunswick, NJ, 2004, pp. 89–126; Report to the President's Council on Bioethics, *Beyond Therapy: Biotechnology and the Pursuit of Happiness*, Dana Press, New York, 2004; Zimmer, C, *Soul Made Flesh: The Discovery of the Brain – and How It Changed the World*, Heinemann, 2004.

Chapter 2

1 I have written about this in more detail in my book *Lifelines: Biology, Freedom, Determinism*, Penguin, Harmondsworth, 1997, 2nd ed. Vintage, forthcoming.

2 Bada, JL and Lazcano, A, 'Prebiotic soup – revisiting the Miller experiment', *Science*, 300, 2003, 745–6.

3 A hypothesis even speculated about by Darwin.

4 Crick, FHC, *Life Itself: Its Origin and Nature*, Macdonald, 1981.

5 Those familiar with the origin of life literature will recognise that this is a modern version of an idea advanced in the 1930s by the Soviet biochemist Alexander Oparin and the British geneticist JBS Haldane.

6 Kauffman, S, *At Home in the Universe: The Search for Laws of Complexity*, Viking, 1995.

7 Daniel Dennett calls natural selection 'a universal acid' as it is a logical consequence of the postulates concerning semi-faithful over-production of offspring; Dennett, D, *Darwin's Dangerous Idea: Evolution and the Meanings of Life*, Allen Lane, 1995.

8 *Mea culpa* – by me too in earlier days.

9 Williams, RJP and Frausto da Silva, JJR, *The Natural Selection of the Chemical Elements*, Oxford University Press, Oxford, 1996.

10 Dawkins, R and Krebs, JR, 'Arms races between and within species', *Proceedings of the Royal Society of London,* B, 205, 1979, 489–511.

11 Koshland, DE Jr, 'The Seven Pillars of Life', *Science*, 295, 2002, 2215–16.

12 Bourne, HR and Weiner, O, 'A chemical compass', *Nature*, 419, 2002, 21.

13 Damasio has argued this in a series of books, most recently Damasio, AR, *Looking for Spinoza: Joy, Sorrow and the Feeling Brain*, Heinemann, 2003.

14 Holmes, FL, *Claude Bernard and Animal Chemistry*, Harvard University Press, Cambridge, Mass, 1974.

15 Gould, Stephen Jay, *The Structure of Evolutionary Theory*, Harvard University Press, Cambridge, Mass, 2002.

16 Brown, Andrew, *In the Beginning Was the Worm: Finding the Secrets of Life in a Tiny Hermaphrodite*, Simon and Schuster, New York, 2003.

17 de Bono, M, Tobin, DM, Davis, MW, Avery, L and Bargmann, C, 'Social feeding in *C. elegans* is induced by neurons that detect aversive stimuli', *Nature*, 419, 2002, 899–903.

18 DeZazzo, J and Tully, T, 'Dissection of memory formation: from behavioural pharmacology to molecular genetics', *Trends in Neuroscience*, 18, 1995, 212–17.

19 Gallistel, R, *The Organisation of Learning*, Bradford Books, New York, 1990.

20 Heisenberg, M, 'Mushroom body memory: from maps to models', *Nature Reviews Neuroscience*, 4, 2003, 266–75.

21 Eric Kandel shared the 2000 Nobel prize for physiology and medicine.

22 O'Keefe, J and Nadel, L, *The Hippocampus as a Cognitive Map*, Oxford University Press, Oxford, 1978.

23 Zeki, S, *A Vision of the Brain*, Blackwell, Oxford, 1993.

24 Deacon, T, *The Symbolic Species: The Co-evolution of Language and the Human Brain*, Allen Lane, The Penguin Press, Harmondsworth, 1997.

25 Finlay, BL and Darlington RB, 'Linked regularities in the development and evolution of the human brain', *Science*, 268, 1995, 1575–84.
26 Kaas, JH and Collins, CE, 'Evolving ideas of brain evolution', *Nature*, 411, 2001, 141–2.
27 Macphail, EM and Bolhuis, JJ, 'The evolution of intelligence: adaptive specialisations *versus* general process', *Biological Reviews*, 76, 2001, 341–64.
28 Churchland, PS and Sejnowski, TJ, *The Computational Brain*, MIT Press, Cambridge, Mass, 1992.
29 Damasio, AR, *Descartes' Error: Emotion, Reason and the Human Brain*, Putnam, New York, 1994.
30 Hilary Rose has gone further and insists, for humans, on 'Amo ergo sum'. Rose, HA, 'Changing constructions of consciousness', *Journal of Consciousness Studies*, 11–12, 1999, 249–56.

Chapter 3

1 Gould, SJ, *Ontogeny and Phylogeny*, Harvard University Press, Cambridge, Mass, 1977.
2 Ridley, M, *Nature via Nurture: Genes, Experience and What Makes Us Human*, Fourth Estate, 2003.
3 Rose, S, *Lifelines: Biology, Freedom, Determinism*, Penguin, Harmondsworth, 1997, 2nd ed. Vintage, forthcoming.
4 Maturana, HR and Varela, FJ, *The Tree of Knowledge: The Biological Roots of Human Understanding*, Shambhala, Boston, 1998.
5 Oyama, S, *The Ontogeny of Information: Developmental Systems and Evolution* (2nd ed.), Duke University Press, Durham, NC, 2000.
6 Warnock, M, *A Question of Life: Warnock Report on Human Fertilization and Embryology,*, Blackwell, Oxford, 1985.
7 Parnavelas, J, 'The human brain, 100 billion connected cells', in Rose, S, ed., *From Brains to Consciousness*, Allen Lane, The Penguin Press, 1998, pp. 18–32.
8 Levi-Montalcini, R, *In Praise of Imperfection: My Life and Work*, Basic Books, New York, 1988.
9 Lewis Wolpert provided the general model for this type of pattern-forming development many years ago, with what he called the 'French flag model', subsequently refined by Brian Goodwin who pointed out that, rather than a continuous gradient, one that pulsed over time provided better three-dimensional control. See his book *Temporal*

Organisation in Cells, Academic Press, New York, 1963.

10 Zeki, S, *A Vision of the Brain*, Blackwell, Oxford, 1994.

11 Purves, D, *Neural Activity and the Growth of the Brain*, Cambridge University Press, Cambridge, 1994.

12 Cohen-Cory, S, 'The developing synapse: construction and modulation of synaptic structures and circuits', *Science,* 298, 2002, 770–6.

13 Edelman, G, *Neural Darwinism,* Basic Books, New York, 1987.

14 See for instance Fausto Stirling, A, *Myths of Gender: Biological Theories about Women and Men*, Basic Books, New York, 1992; Baron Cohen, S, *The Essential Difference: Men, Women and the Extreme Male Brain*, Allen Lane, The Penguin Press, 2003; Jones, S, *Y: The Descent of Man*, Little, Brown, 2002.

15 Roughgarden, J, *Evolution's Rainbow: Diversity, Gender and Sexuality in Nature and People*, University of California Press, Berkeley, CA, 2004.

16 Keverne, EB, 'Genomic imprinting in the brain', *Current Opinion in Neurobiology*, 7, 1997, 463–8; Ferguson-Smith, AC and Surani, MA, 'Imprinting and the epigenetic asymmetry between parental genomes', *Science*, 293, 2001, 1086–93.

Chapter 4

1 For instance in Rose, S, *The Conscious Brain*, Penguin, Harmondsworth, 1973.

2 For instance Churchland, PS and Sejnowski, TJ, *The Computational Brain*, MIT Press, Cambridge, Mass, 1992.

3 Bekoff, M, 'Animal reflections', *Nature*, 419, 2002, 255.

4 For a detailed critique, see Rose, H and Rose, S, eds, *Alas, Poor Darwin: Arguments against Evolutionary Psychology*, Cape, 2000.

5 Proctor, RN, 'Three roots of human recency', *Current Anthropology*, 44, 2003, 213–39.

6 Balter, M, 'What made humans modern?', *Science*, 295, 2002, 1219–25.

7 White, TD, Asfaw, B, DeGusta, D, Gilbert, H, Richards, GD, Suwa, G and Howell, FC, 'Pleistocene *Homo sapiens* from middle Awash, Ethiopia', *Nature*, 423, 2003, 742–7.

8 Carroll, SB, 'Genetics and the making of *Homo sapiens'*, *Nature*, 422, 2003, 849–57.

9 Bromhall, C, *The Eternal Child: An Explosive New Theory of Human Origins and Behaviours*, Ebury, 2003.

10 Cartmill, M, 'Men behaving childishly', *Times Literary Supplement*, 5223, 2003, 28.

11 Such measurements can of course also be very inaccurate – as pointed out by Stephen J Gould in his book *The Mismeasure of Man*, Norton, New York, 1981.

12 Enard, W and thirteen others, 'Intra- and inter-specific variation in primate gene expression patterns', *Science*, 296, 2002, 340–2.

13 Cavalli-Sforza, LL, *Genes, Peoples and Languages*, Allen Lane, The Penguin Press, 2000.

14 Runciman, WG, *The Social Animal*, HarperCollins, 1998.

15 Boyd, R and Richardson, PJ, *Culture and the Evolutionary Process*, University of Chicago Press, Chicago, 1998.

16 For an absurdly enthusiastic development of this, see Blackmore, S, *The Meme Machine*, Oxford University Press, Oxford, 1999; for its demolition, see Mary Midgley's chapter in Rose, H and Rose, S, eds, *Alas, Poor Darwin*, op. cit., pp. 67–84.

17 Ofek, H, *Second Nature: Economic Origins of Human Evolution*, Cambridge University Press, Cambridge, 2001.

18 See for instance Cleeremans, A, ed., *The Unity of Consciousness: Binding, Integration and Dissociation*, Oxford University Press, Oxford, 2003.

19 For instance Gould, op. cit.; Rose, SPR, Lewontin, RC and Kamin, L, *Not in Our Genes*, Allen Lane, The Penguin Press, Harmondsworth, 1984.

20 Rose, SPR in Rose, H and Rose, S, eds, *Alas, Poor Darwin*, op. cit., pp. 247–65, from which part of this argument has been extracted.

21 See for example Ridley, M, *Genome: The Autobiography of a Species in 23 Chapters*, Fourth Estate, 1999.

22 Gazzaniga, M, *The Social Brain*, Basic Books, New York, 1985.

23 Rose, H in *Alas, Poor Darwin*, op. cit., pp. 106–28.

24 Hamilton, WD, 'The genetical evolution of social behaviour, I and II', *Journal of Theoretical Biology*, 7, 1964, 1–32.

25 Trivers, RL, 'The evolution of reciprocal altruism', *Quarterly Review of Biology*, 46, 1971, 35–57.

26 Betzig, L, ed., *Human Nature: A Critical Reader*, Oxford University Press, New York, 1997, Introduction.

27 Thornhill, R and Palmer, CT, *A Natural History of Rape: Biological Bases of Sexual Coercion*, MIT Press, Cambridge, Mass, 2000.

28 Haraway, D, *Primate Visions: Gender, Race and Nature in the World of*

Modern Science, Routledge, 1989.

29 Eldredge, N, *Time Frames*, Simon and Schuster, New York, 1985.

30 Jones, S, *Almost Like a Whale*, Doubleday, 1999, p. 242.

31 For example, see Dehaene, S, *The Number Sense*, Oxford University Press, New York, 1997.

32 Fodor, J, *The Modularity of Mind*, Bradford Books, MIT Press, Cambridge, Mass, 1983.

33 Mithen, S, *The Prehistory of the Mind*, Thames and Hudson, 1996.

34 Fodor, J, *The Mind Doesn't Work That Way: The Scope and Limits of Computational Psychology*, Bradford Books, MIT Press, Cambridge, Mass, 2000.

35 See for example Rose, S, *The Making of Memory*, Bantam, 1992, 2nd ed. Vintage, 2003; Rose, S, ed., *From Brains to Consciousness: Essays on the New Sciences of the Mind*, Allen Lane, The Penguin Press, 1998; Freeman, W, *How Brains Make up Their Minds,* Weidenfeld and Nicolson, 1999.

36 For example, Damasio, AR, *Descartes' Error: Emotion, Reason and the Human Brain,* Putnam, New York, 1994, and *The Feeling of What Happens,* Heinemann, 1999; LeDoux, J, *The Emotional Brain: The Mysterious Underpinnings of Emotional Life,* Simon and Schuster, New York, 1996.

37 Betzig, op. cit.

38 For a discussion of this claim as to the appeal of Bayswater Road art, see Jencks, C, 'EP, phone home', in *Alas, Poor Darwin,* op. cit., pp. 28–47.

39 Schama, S, *Landscape and Memory*, HarperCollins, 1995.

40 For a discussion of the US violence initiative see for instance Breggin, PR and Breggin, GR, *A Biomedical Program for Urban Violence Control in the US: The Dangers of Psychiatric Social Control,* Center for the Study of Psychiatry (mimeo), 1994.

41 See Betzig, op. cit.

42 Wilson, EO, *Consilience: The Unity of Knowledge,* Little, Brown, 1998.

43 Hauser, M, *Wild Minds: What Animals Really Think,* Penguin, 2001.

44 Hauser, MD, Chomsky, N and Fitch, WT, 'The faculty of language: what is it, who has it and how did it evolve?', *Science*, 298, 2002, 1569–79.

45 Sinha, C, 'Biology, culture and the emergence and elaboration of

symbolisation', in Saleemi, A, Gjedde, A and Bohn, O-S, eds, *In Search of a Language for the Mind-Brain: Can the Multiple Perspectives be Unified?* Aarhus University Press, Aarhus, forthcoming.

46 Deacon, T, *The Symbolic Species,* Allen Lane, The Penguin Press, Harmondsworth, 1997.

47 Savage-Rumbaugh, S, Shanker, SG and Taylor, TJ, *Apes, Language and the Human Mind,* Oxford University Press, Oxford, 1998.

48 Lai, CSL, Fisher, SE, Hurst, JA, Vargha-Khadem, F and Monaco, AP, 'A forkhead gene is mutated in a severe speech and language disorder', *Nature,* 413, 2001, 519–23.

49 Enard, W and seven others, 'Molecular evolution of FOXP2, a gene involved in speech and language', *Nature,* 418, 2002, 869–72.

50 Savage-Rumbaugh, ES and Lewin, R, *Kanzi: The Ape at the Brink of the Human Mind,* John Wiley, New York, 1994.

51 Ghazanfar, AA and Logothetis, NK, 'Facial expressions linked to monkey calls', *Nature,* 423, 2003, 937.

52 Chomsky, N, *Syntactic Structures,* Mouton, The Hague, 1957.

53 Pinker, S, *The Language Instinct: The New Science of Language and Mind,* Allen Lane, The Penguin Press, Harmondsworth, 1994.

54 Donald, M, *A Mind so Rare: The Evolution of Human Consciousness,* Norton, New York, 2002.

Chapter 5

1 Bateson, PPG, 'Taking the stink out of instinct', in Rose, H and Rose, S, eds, *Alas, Poor Darwin: Arguments against Evolutionary Psychology*, Cape, 2000, pp. 157–73.

2 Ingold, T, 'Evolving skills', in *Alas, Poor Darwin*, op. cit., pp. 225–46.

3 Geber, M, and Dean, RFA, *Courier*, 6 (3), 1956.

4 Eliot, L, *Early Intelligence: How the Brain and Mind Develop in the First Five Years of Life*, Penguin, 1999.

5 Crick, FHC, *The Astonishing Hypothesis: The Scientific Search for the Soul*, Simon and Schuster, 1994.

6 Bruce, V and Young, A, *In the Eye of the Beholder*, Oxford University Press, Oxford, 1998, p. 280, quoted by Bateson, op. cit.

7 Johnson, M and Morton, J, *Biology and Cognitive Development: The Case of Face Recognition*, Blackwell, Oxford, 1991.

8 Trehub, S, 'The developmental origins of musicality', *Nature Neuroscience*, 6, 2003, 669–73.

9 Hubel, DH, *Eye, Brain and Vision*, Scientific American Library, New York, 1988.

10 Johnson, MH, 'Brain and cognitive development in infancy', *Current Opinion in Neurobiology*, 4, 1994, 218–25.

11 Kendrick, KM, da Costa, AP, Leigh, AE, Hinton, MR and Peirce, JW, 'Sheep don't forget a face', *Nature*, 414, 2001, 165–6.

12 Schaal, B, Coureaud, G, Langlois, D, Ginies, C, Seemon, E and Perrier, G, 'Chemical and behavioural characteristics of the rabbit mammary pheromone', *Nature*, 424, 2003, 68–72.

13 Pascalis, O, de Haan, M and Nelson, CA, 'Is face processing species-specific during the first years of life?', *Science*, 296, 2002, 1321–3.

14 Golby, AJ, Gabrieli, JDE, Chiao, JY and Eberhardt, JL, 'Differential responses in the fusiform region to same-race and other-race faces', *Nature Neuroscience*, 4, 2001, 845–50.

15 Liston, C and Kagan, J, 'Memory enhancement in early childhood', *Nature*, 419, 2002, 896.

16 I have drawn this time sequence from Eliot, op. cit.

17 Rivera-Gaxiola, M, Csibra, G, Johnson, MH and Karmiloff-Smith, A, 'Electrophysiological correlates of cross-linguistic speech perception in native English speakers', *Behavioural Brain Research*, 111, 2000, 13–23.

18 Pinker S, *The Language Instinct: The New Science of Language and Mind*, Allen Lane, The Penguin Press, Harmondsworth, 1994.

19 Pinker, S, 'Talk of genetics and vice versa', *Nature*, 413, 2001, 465–6.

20 Karmiloff-Smith, A, 'Bates' emergentist theory and its relevance to the exploration of genotype/phenotype interactions', in Sobin, D and Tomasello, M, *Beyond Nature–Nurture: Essays in Honor of Elizabeth Bates*, Erlbaum, Mahwah, NJ, forthcoming.

21 Karmiloff-Smith, A, 'Why babies' brains are not Swiss army knives', in *Alas, Poor Darwin* op. cit., pp. 144–56.

22 Wexler, K, 'The development of inflection in a biologically-based theory of language acquisition', in Rice, M, ed., *Towards a Genetics of Language*, Erlbaum, Mahwah, NJ, 1996, 113–44, quoted by Karmiloff-Smith, op. cit.

23 Smith, N and Tsimpli, I, *The Mind of a Savant: Language Learning and Modularity*, Blackwell, Oxford, 1995, quoted by Karmiloff-Smith, op. cit.

24 Turkeltaub, PE, Gareau, L, Flowers, DL, Zeffiro, TA and Eden, GF,

'Development of neural mechanisms for reading', *Nature Neuroscience*, 6, 2003, 767–73.

25 Bateson, op. cit.

26 Baron Cohen, S, *The Essential Difference: Men, Women and the Extreme Male Brain*, Allen Lane, The Penguin Press, 2003.

27 Hobson, P, *The Cradle of Thought*, Macmillan, 2001.

Chapter 6

1 Gould, SJ and Lewontin, RC, 'The spandrels of San Marco and the Panglossian paradigm: a critique of the adaptationist programme', *Proceedings of the Royal Society*, B, 205, 1979, 581–98.

2 Churchland, PM, *Brain-wise: Studies in Neurophilosophy*, Bradford Books, MIT Press, Cambridge, Mass, 2002.

3 Boden M, 'Does artificial intelligence need artificial brains?', in Rose, S and Appignanesi, L, eds, *Science and Beyond*, Blackwell, Oxford, 1986, pp. 103–14.

4 Aleksander, I, *How to Build a Mind*, Weidenfeld and Nicolson, 2001.

5 Hauser, M, *Wild Minds: What Animals Really Think*, Penguin, 2001.

6 Damasio, AR, *The Feeling of What Happens*, Heinemann, 1999.

7 Rose, S, *The Making of Memory*, Bantam, 1992, 2nd ed. Vintage, 2003.

8 Penrose, R, *The Emperor's New Mind: Concerning Computers, Minds and the Laws of Physics*, Oxford University Press, Oxford, 1989.

9 Svendsen, CN, 'The amazing astrocyte', *Nature*, 417, 2002, 29–31.

10 Acebes, A and Ferrus, A, 'Cellular and molecular features of axon collaterals and dendrites', *Trends in Neuroscience*, 23, 2000, 557–65.

11 Hausser, M, Spruston, N and Stuart, GJ, 'Diversity and dynamics of dendritic signalling', *Science*, 290, 2000, 739–44.

12 Purves, D, *Neural Activity and the Growth of the Brain*, Cambridge University Press, Cambridge, 1994.

13 Trachtenberg, JT, Chen, BE, Knott, GW, Feng, G, Sanes, JR, Welker, E and Svoboda, K, 'Long-term in vivo imaging of experience-dependent synaptic plasticity in adult cortex', *Nature*, 420, 2002, 788–94.

14 Eccles, JC, *Facing Reality*, Springer, New York, 1970.

15 Novartis Symposium 213, *The Limits of Reductionism in Biology*, Wiley, New York, 1998.

16 Ledoux, J, *Synaptic Self: How Our Brains Become Who We Are*, Viking Penguin, 2002.

17 Eccles, JC, Itoh, M and Szentagothai, J, *The Cerebellum as a Neuronal Machine*, Springer, New York, 1967.

18 Attwell, PJE, Cooke, SF and Yeo, CH, 'Cerebellar function in consolidation of a motor memory', *Neuron*, 34, 2002, 1011–20.

19 Weiskrantz, L, Warrington, EK, Sanders, MD and Marshall, J, 'Visual capacity in the hemianopic field following a restricted cortical ablation', *Brain*, 97, 1974, 709–28.

20 Zeki, S, *A Vision of the Brain*, Blackwell, Oxford, 1974.

21 Dennett, DC, *Consciousness Explained*, Allen Lane, 1991.

22 Wall, P, *Pain: The Science of Suffering*, Weidenfeld and Nicolson, 2000.

23 Brautigam, S, Stins, JF, Rose, SPR, Swithenby, SJ and Ambler, T, 'Magnetoencephalographic signals identify stages in real-life decision processes', *Neural Plasticity*, 8, 2001, 241–53.

24 Damasio, AR, op. cit.

25 Elbert, T and Heim, S, 'A light and a dark side', *Nature*, 411, 2001, 139.

26 Ramachandran, VS and Blakeslee, S, *Phantoms in the Brain: Human Nature and the Architecture of the Mind*, Fourth Estate, 1998.

27 Walter, WG, *The Living Brain*, Pelican Books, Harmondsworth, 1961.

28 Singer, W, 'Consciousness from a neurobiological perspective', in Rose, S, ed., *Brains to Consciousness: Essays on the New Sciences of the Mind*, Allen Lane, The Penguin Press, 1998, pp. 228–45.

29 Freeman, WJ, *How Brains Make up Their Minds*, Weidenfeld and Nicolson, 1999.

30 Rose, S, *The Making of Memory*, op. cit.

31 Hebb, DO, *The Organisation of Behavior*, Wiley, New York, 1949.

32 Milner, B, Corkin, S, and Teuber, HL, 'Further analysis of the hippocampal amnestic syndrome: 14 year follow-up study of HM', *Neuropsychologia*, 6, 1968, 215–34.

33 For example, Anokhin, KV, Tiunova, AA and Rose, SPR, 'Reminder effects – reconsolidation or retrieval deficit? Pharmacological dissection following reminder for a passive avoidance task in young chicks', *European Journal of Neuroscience* 15, 2002, 1759–65.

34 Ojemann, GA, Schenfield-McNeill, J and Corina, DP, 'Anatomical subdivisions in human temporal cortex neuronal activity related to verbal memory', *Nature Neuroscience*, 5, 2002, 64–71.

35 Kim, JJ and Baxter, MG, 'Multiple memory systems: the whole does not equal the sum of its parts', *Trends in Neuroscience*, 24, 2001, 324–30.

36 Buckner, RL and Wheeler, ME, 'The cognitive neuroscience of remembering', *Nature Reviews Neuroscience*, 2, 2001, 624–34.

37 Interview with Endel Tulving, *Journal of Cognitive Neuroscience*, 3, 1991, 89.

38 Anokhin, PK, *Biology and Neurophysiology of the Conditional Reflex and its Role in Adaptive Behaviour*, Pergamon Press, Oxford, 1974.

39 Argued for instance by Churchland, PS and Churchland, PM, 'Neural worlds and real worlds', *Nature Reviews Neuroscience* 3, 2002, 903–7.

40 O'Keefe, J and Nadel, L, *The Hippocampus as a Cognitive Map*, Oxford University Press, Oxford, 1978.

41 Gaffan, D, 'Against memory systems', *Philosophical Transactions of the Royal Society*, B, 357, 2002, 1111–21.

42 Rose, S, *The Conscious Brain*, Weidenfeld and Nicolson, 1973.

43 McGaugh, JL, *Memory and Emotion: The Making of Lasting Memories*, Weidenfeld and Nicolson, 2003.

44 McEwen, BS, Schmeck, HM and Kybiuk, L, *The Hostage Brain*, Rockefeller University Press, New York, 1998.

45 Johnston, ANB and Rose, SPR, 'Isolation-stress induced facilitation of passive avoidance memory in the day-old chick', *Behavioral Neuroscience*, 112, 1998, 1–8; Sandi, C and Rose, SPR, 'Corticosteroid receptor antagonists are amnestic for passive avoidance learning in day-old chicks', *European Journal of Neuroscience*, 6, 1994, 1292–7.

46 Damasio, AR, in a series of books, most recently *Looking for Spinoza: Joy, Sorrow and the Feeling Brain*, Heinemann, 2003.

47 Tononi, G, 'Consciousness, differentiated and integrated', in Cleeremans, A, ed., *The Unity of Consciousness: Binding, Integration and Dissociation*, Oxford University Press, Oxford 2003, pp. 253–65.

48 Lodge, D, *Consciousness and the Novel*, Secker and Warburg, 2002.

49 Rose, H, 'Consciousness and the Limits of Neurobiology', in Rees, D and Rose, S, eds, *The New Brain Sciences: Prospects and Perils*, Cambridge University Press, Cambridge, 2004, pp. 59–70.

50 Adolphs, R, 'The cognitive neuroscience of human social behaviour', *Nature Reviews Neuroscience*, 4, 2003, 165–77.

51 Frith, CD and Wolpert, DM, eds, 'Decoding, imitating and influencing the actions of others: the mechanisms of social interactions', *Philosophical Transactions of the Royal Society*, B, 358, 2003, 429–602.

52 Blakemore, S-J and Decety, J, 'From the perception of action to the understanding of intention', *Nature Reviews Neuroscience*, 2, 2001,

561–7; Ramnani, N and Miall, RC, 'A system in the human brain for predicting the actions of others', *Nature Neuroscience*, 7, 2004, 85–90.

53 Singer, T, Seymour, B, O'Doherty, J, Kaube, H, Dolan, RJ and Frith, CD, 'Empathy for pain involved the affective but not sensory components of pain', *Science*, 303, 2004, 1157–62.

54 Siegal, M and Varley, R, 'Neural systems involved in "theory of mind"', *Nature Reviews Neuroscience*, 3, 2002, 463–71.

Chapter 7

1 Langreth, R, 'Viagra for the brain', *Forbes* magazine, 4 February 2002.

2 Hayflick, L, 'Origins of longevity', in Warner, HR et al., eds, *Modern Biological Theories of Aging*, Raven Press, New York, 1987, pp. 21–34.

3 Whalley, L, *The Ageing Brain*, Weidenfeld and Nicolson, 2001; Kirkwood, T, *Time of Our Lives: Why Ageing Is Neither Inevitable Nor Necessary*, Phoenix, 2000.

4 Hayflick, op. cit.

5 Johnson, SA and Finch, CE, 'Changes in gene expression during brain aging: a survey', in Schneider, E and Rowe, J, eds, *Handbook of the Biology of Aging*, Academic Press, New York, 1996, pp. 300–27.

6 Barnes, CA, 'Normal aging: regionally specific changes in hippocampal synaptic transmission', *Trends in Neuroscience*, 17, 1994, 13–18.

7 McEchron, MD, Weible, AP and Disterhoft, JF, 'Aging and learning specific alterations in single neuron ensembles in the CA1 area of the hippocampus during rabbit trace eyeblink conditioning', *Journal of Neurophysiology*, 86, 2001, 1839–57.

8 Rabbitt, P, 'Does it all go together when it goes?', *Quarterly Journal of Experimental Psychology*, 46A, 1993, 354–85.

9 Quoted by Cohen, G, 'Memory and learning in normal ageing', in Woods, RT, ed., *Handbook of the Clinical Psychology of Ageing*, Wiley, New York, 1996, pp. 43–58.

10 Salthouse, TA, 'Mediation of adult age differences in cognition by reductions in working memory and speed of processing', *Psychological Science*, 2, 1991, 179–83.

11 Youdim, MBH and Riederer, P, 'Understanding Parkinson's Disease', *Scientific American*, January 1997, pp. 38–45.

12 Sacks, O, *Awakenings*, Duckworth, 1973.

13 Rees, D and Rose, S, eds, *The New Brain Sciences: Prospects and Perils*, Cambridge University Press, Cambridge, 2004.

14 Strittmater, WJ and Roses, AD, 'Apolipoprotein E and Alzheimer's disease', *Annual Review of Neuroscience*, 19, 1996, 53–77.

15 Schneider, LS and Finch, CE, 'Can estrogens prevent neurodegeneration?', *Drugs and Aging*, 11, 1997, 87–95.

16 Selkoe, DJ, 'Normal and abnormal biology of the beta-amyloid precursor protein', *Annual Review of Neuroscience*, 17, 1994, 489–517.

17 Mileusnic, R, Lancashire, CL, Johnson, ANB and Rose, SPR, 'APP is required during an early phase of memory formation', *European Journal of Neuroscience*, 12, 2000, 4487–95.

18 Rose, S, *The Making of Memory*, Bantam, 1992, 2nd ed. Vintage, 2003.

Chapter 8

1 For a thoroughly reductionist biologist's attempt to come to terms with his own depression, see Wolpert, L, *Malignant Sadness: The Anatomy of Depression*, Faber and Faber, 2001.

2 Pinker, S, *How the Mind Works*, Allen Lane, 1997.

3 Dawkins, R, *The Selfish Gene*, Oxford University Press, Oxford, 1976.

4 Wilson, EO, *Consilience: The Unity of Knowledge*, Little, Brown, 1998.

5 Miller, G, *The Mating Mind*, Heinemann, 2000.

6 Jencks, C, 'EP, phone home', in Rose, H and Rose, S, eds, *Alas, Poor Darwin: Arguments against Evolutionary Psychology*, Cape, 2000, pp. 33–54.

7 Rose, H and Rose, S, *Science and Society*, Allen Lane, 1969.

8 Goodwin, B, *How the Leopard Changed its Spots*, Weidenfeld and Nicolson, 1994.

9 Needham, J and successors, *Science and Civilisation in China*, Cambridge University Press, Cambridge, continuing series, 1956– .

10 Zimmer, C, *Soul Made Flesh: The Discovery of the Brain – and How It Changed the World*, Heinemann, 2004.

11 Finger, S, *The Origins of Neuroscience*, Oxford University Press, New York, 1994.

12 Sorabji, R, *Aristotle on Memory*, Brown University Press, Providence, 1972.

13 St Augustine, *Confessions*, translated by RS Pine-Coffin, Penguin, Harmondsworth, 1961.

14 Ibid., p. 214.

15 Ibid., p. 215.

16 Ibid., pp. 216–20.

17 Descartes, R, *Discourse on Method and Related Writings*, Penguin, 1999 (1637).

18 Wall, P, *Pain: The Science of Suffering*, Orion, 2000.

19 Zimmer, op. cit.

20 Israel, JI, *Radical Enlightenment: Philosophy and the Making of Modernity 1650–1750*, Oxford University Press, Oxford, 2001.

21 McGinn, C, *The Mysterious Flame: Conscious Minds in a Material World*, Basic Books, New York, 1999.

22 Quoted in Finger, op. cit., p. 33.

23 Lashley, KS, 'In search of the engram', *Symposia of the Society for Experimental Biology*, 4, 1950, 553–61.

24 Harrington, A, *Medicine, Mind and the Double Brain: A Study in Nineteenth Century Thought*, Princeton University Press, Princeton, NJ, 1987.

25 Ibid., p. 145.

26 Abraham, C, *Possessing Genius: The Bizarre Odyssey of Einstein's Brain*, Icon Books, Cambridge, 2004.

27 LeVay, S, *Queer Science: The Use and Abuse of Research into Homosexuality*, MIT Press, Cambridge, Mass, 1996.

28 Morgan, M, *The Space Between Our Ears: How the Brain Represents Visual Space*, Weidenfeld and Nicolson, 2003.

29 Pinker, S, *How the Mind Works*, Allen Lane, 1997.

30 Standing, L, 'Remembering ten thousand pictures', *Quarterly Journal of Experimental Psychology*, 25, 1973, 207–22.

31 Yates, F, *The Art of Memory*, Penguin, Harmondsworth, 1966.

32 Squire, LR, *Memory and Brain*, Oxford University Press, New York, 1987; see also Squire, LR, Clark, RE and Knowlton, BJ, 'Retrograde amnesia', *Hippocampus*, 11, 2001, 50–5.

33 Dudai, Y, *The Neurobiology of Learning and Memory: Concepts, Findings, Trends*, Oxford University Press, Oxford, 1989.

34 Baddeley, A, *Essentials of Human Memory*, Psychology Press, Sussex, 1999.

35 Chalmers, D, *The Conscious Mind: In Search of a Fundamental Theory*, Oxford University Press, New York, 1996.

36 le Bihan, D, Claude Bernard Lecture, The Royal Society, 2004.

37 Blakemore, S-J and Decety, J, 'From the perception of action to the understanding of intention', *Nature Reviews Neuroscience*, 2, 2001, 561–7.

38 Dehaene, S, *The Number Sense: How the Mind Creates Mathematics*, Oxford University Press, New York, 1997.

39 Schacter, DL, *The Cognitive Psychology of False Memories*, Psychology Press, Sussex, 1999.

Chapter 9

1 Kaptchuk, TJ, *Chinese Medicine: The Web that Has No Weaver*, Rider, 1983.

2 Sugarman, PA and Crauford, D, 'Schizophrenia and the Afro-Caribbean community', *British Journal of Psychiatry*, 164, 1994, 474 80. See also Leff, J, *The Unbalanced Mind*, Weidenfeld and Nicolson, 2001.

3 Laing, RD, *The Divided Self*, Penguin, Harmondsworth, 1965.

4 Warner, R, *Recovery from Schizophrenia: Psychiatry and Political Economy*, Routledge and Kegan Paul, 1985.

5 Rosenhan DL, 'On being sane in insane places', *Science*, 179, 1973, 250–8.

6 Warner, R, op. cit.

7 Bloch, S and Reddaway, P, *Russia's Political Hospitals: The Abuse of Psychiatry in the Soviet Union*, Gollancz, 1977; see also Medvedev, ZA and Medvedev, RA, *A Question of Madness*, Macmillan, 1971.

8 Slater, L, *Opening Skinner's Box: Great Psychological Experiments of the 20th Century*, Bloomsbury, 2004 (as extracted in the *Guardian*, 31 January 2004, pp. 14–23).

9 Bentall, RP, *Madness Explained: Psychosis and Human Nature*, Allen Lane, 2003.

10 Watson, JB, *Behaviourism*, Transaction Press, New Brunswick, NJ, 1924.

11 Pinker, S, *The Blank Slate: The Modern Denial of Human Nature*, Allen Lane, 2002.

12 Skinner, BF, *Walden Two*, Macmillan, 1976; *Beyond Freedom and Dignity*, Cape, 1972.

13 Shutts, D, *Lobotomy: Resort to the Knife*, Van Nostrand Reinhold, New York, 1982.

14 Mark, VH and Ervin, FR, *Violence and the Brain*, Harper & Row, New York, 1970.

15 Opton, N, correspondence circulated at Winter Conference on Brain Research, Vail, Colorado, 1973.

16 Delgado, JMR, *Physical Control of the Mind*, Harper & Row, 1971.

17 Curran, D and Patridge, M, *Psychological Medicine*, 6th edition, Livingstone, 1969.

18 *British National Formulary*, 45, 2003, p. 171.

19 *Guardian*, 11 December 2003, p. 13.

20 Bignami, G, 'Disease models and reductionist thinking in the biomedical sciences', in Rose, S, ed., *Against Biological Determinism*, Allison and Busby, 1982, pp. 94–110.

21 Barondes, S, *Better than Prozac*, Oxford University Press, New York, 2003, p. 34; the quotation is from the psychiatrist Roland Kuhn.

22 Willis, S, 'The influence of psychotherapy and depression on platelet imipramine and paroxetine binding', thesis, Open University, Milton Keynes, 1992.

23 Dennis, J, 'Antidepressant market forecasts, 2003–2008', *Visiongain.com* report, 2003.

24 Kramer, PD, *Listening to Prozac: A Psychiatrist Explores Anti-depressant Drugs and the Remaking of Self*, Fourth Estate, 1993.

25 Cornwell, J, *The Power to Harm: Mind, Medicine and Murder on Trial*, Viking, 1996.

26 Breggin, PR, *Talking Back to Prozac*, St Martin's Press, New York, 1995.

27 Healy, D, *Let Them Eat Prozac*, Lorimer, Toronto, 2003.

Chapter 10

1 Rees, D and Rose, SPR, eds, *The New Brain Sciences: Prospects and Perils*, Cambridge University Press, Cambridge, 2004.

2 This section is based on an earlier article of mine; Rose, SPR, 'Smart drugs: will they work, are they ethical, will they be legal?', *Nature Reviews Neuroscience*, 3, 2002, 975–9.

3 Tang, YP and seven others, 'Enhancement of learning and memory in mice', *Nature*, 401, 1999, 63–9.

4 Langreth, R, 'Viagra for the brain', *Forbes* magazine, 4 February 2002.

5 Giurgia, C, 'Vers une pharmacologie de l'activité integrative du cerveau. Tentative du concept nootrope en psychopharmacologie', *Actualité, Pharmacologie*, 1972. Thanks to Susan Sara for this quote.

6 Dean, W, and Morgenthaler, J, *Smart Drugs and Nutrients*, B & J Publications, Santa Cruz, CA, 1991.

7 Bolla, KI, Lindgren, KN, Bonaccorsy, C and Bleecker, ML, 'Memory complaints in older adults: fact or fiction?', *Archives of Neurology*, 48, 1991, 61–5.

8 Yates, F, *The Art of Memory*, Penguin, 1966.

9 Loftus, E, and Ketchum, K, *The Myth of Repressed Memory: False Memories and Allegations of Sexual Abuse*, St Martin's Press, New York, 1994.

10 Luria, AR, *The Mind of a Mnemonist*, Cape, 1969.

11 Borges, JL, *Funes the Memorious*, Calder, 1965.

12 Sandi, C, and Rose, SPR, 'Training-dependent biphasic effects of corticosterone in memory formation for passive avoidance tasks in chicks', *Psychopharmacology*, 133, 1997, 152–60; McGaugh, JL and Roozendaal, B, 'Role of adrenal stress hormones in forming lasting memories in the brain', *Current Opinion in Neurobiology*, 12, 2002, 205–10; Gold, PE, 'Glucose modulation of memory storage processing', *Behavioral and Neural Biology*, 45, 1986, 145–55.

13 Burbach, JPH and de Wied, D, *Brain Functions of Neuropeptides*, Parthenon, Carnforth, 1993.

14 Migues, PV, Johnston, ANB and Rose, SPR, 'Dehydro-epiandrosterone and its sulphate enhance memory retention in day-old chicks', *Neuroscience*, 109, 2001, 243–51; Johnston, ANB and Rose, SPR, 'Memory consolidation in day-old chicks requires BDNF but not NGF or NT-3: an antisense study', *Molecular Brain Research*, 88, 2001, 26–36; Paganini-Hill, A, and Henderson, VW, 'Estrogen deficiency and risk of Alzheimer's disease in women', *American Journal of Epidemiology*, 140, 1994, 256–61; Schneider, LS and Finch, CE, 'Can estrogens prevent neurodegeneration?' *Drugs and Aging*, 11, 1997, 87–95.

15 Bourtchouladze, R and five others, 'Deficient long-term memory in mice with a targeted mutation in the cAMP-responsive element binding protein', *Cell*, 79, 1994, 59–68.

16 Kogan, JH et al., 'Spaced training indices of normal memory in CREB mutant mice', *Current Biology* 7, 1997, 1–11.

17 AD 2000 collaboration group, 'Long-term donepazil treatment in 565 patients with Alzheimer's Disease', *Lancet*, 363, 2004, 2105–15.

18 Sacks, O, *Awakenings*, Duckworth, 1973.

19 Breithaupt, H, and Weigmann, K, 'Manipulating your mind', *EMBO Reports*, 5, 2004, 230–2.

20 McGaugh, JL, 'Enhancing cognitive performance', *Southern Californian Law Review*, 65, 1991, 383–95.

21 Editorial, 'Tetra-tab – cognitive enhancement gone wrong', *Lancet Neurology*, 2, 2003, 151.

22 Quoted by Brown, K, 'New attention to ADHD genes', *Science*, 301, 2003, 160–1.

23 Barkley, RA, Cook, ER Jr, Diamond, A et al., 'International consensus statement on ADHD', *Clinical Child and Family Psychology Review*, 5, 2002, 89–111.

24 Cooper, P, 'Education in the age of Ritalin', in Rees, D and Rose, SPR, eds, *The New Brain Sciences: Prospects and Perils*, Cambridge University Press, Cambridge, 2004, pp. 249–62.

25 Timimi, S, 'ADHD is best understood as a cultural construct', *British Journal of Psychiatry*, 184, 2004, 8–9.

26 Morrison, JR and Stewart, MA, 'A family study of the hyperactive child syndrome', *Biological Psychiatry*, 3, 1971, 189–95.

27 Rose, SPR, Lewontin, RC and Kamin, LJ, *Not in our Genes*, 1984, Penguin, Harmondsworth, Chapter 7.

28 Rose, Lewontin and Kamin, op. cit.; Balaban, E, 'Behaviour genetics: Galen's prophecy or Malpighi's legacy?', in Singh, RS, Krimbas, CB, Paul, DB and Beatty, J, eds, *Thinking about Evolution: Historical, Philosophical, and Political Perspectives*, vol. 2, Cambridge University Press, Cambridge, 2001, pp. 429–66.

29 Rose, Lewontin and Kamin, op. cit.

30 Brown, K, op. cit.

31 Cantwell, DP, 'Drugs and medical intervention', in Rie, HE and Rie, ED, eds, *Handbook of Minimal Brain Dysfunctions*, Wiley, New York, 1980, pp. 596–7.

32 Ellins, J, 'Oh behave!', quoted by Studwell, J, *Financial Times*, 23 January 2004.

33 Breggin, PR, *The Ritalin Factbook*, Perseus, Cambridge, Mass, 2002.

34 Studwell, *Financial Times* interview, op. cit.

35 Healy, D, in Rees and Rose, op. cit., pp. 232–48.

36 The Edinburgh organisation is Overload; its website is Overloadnetwork.org.uk.

Chapter 11

1 Rose, H, 'Gendered genetics in Iceland', *New Genetics and Society*, 20, 2001, 119–38.

2 Report to the President's Council on Bioethics, *Beyond Therapy: Biotechnology and the Pursuit of Happiness*, Dana Press, New York, 2003.

3 Ekman, P, *Emotions Revealed: Recognising Faces and Feelings to Improve*

Communication and Emotional Life, Henry Holt, New York, 2003.

4 Golby, AJ, Gabrieli, JDE, Chiaio, JY and Eberhardt, JL, 'Differential responses in the fusiform region to same-race and other-race faces', *Nature Neuroscience*, 4, 2001, 845–50.

5 Richeson, JA, Baird, AA, Gordon, HL, Heatherton, TF, Wyland, CL, Trawaler, S and Shelton, JN, 'An fMRI investigation of the impact of interracial contact on executive function', *Nature Neuroscience*, 6, 2003, 1323–8.

6 Leighton, N, 'They're reading our minds', *Sunday Times*, 25 January 2004.

7 Pulvermuller, F, 'Words in the brain's language', *Behavioral and Brain Sciences*, 22, 1999, 253–336; Skrandies, W, 'Evoked potential correlates of semantic meaning – a brain mapping study', *Cognitive Brain Research*, 6, 1998, 175–83; Pulvermuller, F, Mohn, B and Schleichert, H, 'Semantic or lexico-syntactic factors: what determines word-class specific activity in the human brain?' *Neuroscience Letters*, 275, 1999, 81–4.

8 Guyatt, DG, 'Some aspects of anti-personnel electromagnetic weapons', International Committee of the Red Cross Symposium: The Medical Profession and the Effects of Weapons, ICRC publication ref. 06681996.

9 Brodeur, P, *The Zapping of America*, Norton, New York, 1977.

10 Papert, S, 'One AI or many?', in Graubert, SR, ed., *The Artificial Intelligence Debate*, MIT Press, Cambridge, Mass, 1988, pp. 3–4.

11 McMurtrey, J, 'Remote behavioral influence technology evidence', circulated paper, 903 N. Calvert St, Baltimore, MD 21202, 2003.

12 Kiyuna, T, Tanigawa, T and Yamazaki, T, Patent #5785653, 'System and method for predicting internal condition of live body', USPTO, granted 7/28/98.

13 Witchalls, C, 'Murder in mind', *Guardian* Life supplement, 25 March 2004.

14 Moir, A and Jessel, D, *A Mind to Crime*, Michael Joseph, 1995.

15 Raine, A, *The Psychopathology of Crime: Criminal Behaviour as a Clinical Disorder*, Academic Press, San Diego, 1993.

16 Rogers, L, 'Secret tests on brains of serial killers', *Sunday Times*, 4 April 2004.

17 E.g., Jones, G, *Social Darwinism and English Thought*, Harvester Press, Sussex, 1980; Dickens, DK, *Eugenics and the Progressives*, Vanderbilt

University Press, Nashville, TN, 1968: Rose, S, Lewontin, RC and Kamin, L, *Not in Our Genes*, Penguin, Harmondsworth, 1984.

18 Rushton, JP, *Race, Evolution and Behaviour: A Life History Perspective*, Transaction Publishers, New Brunswick, NJ, 1995.

19 Balaban, E, 'Behaviour genetics: Galen's prophecy or Malpighi's legacy?' in Singh, RS, Krimbas, CB, Paul, DB and Beatty, J, eds, *Thinking about Evolution: Historical, Philosophical, and Political Perspectives*, vol. 2, Cambridge University Press, Cambridge, 2001, pp. 429–66.

20 Kagan, J, *Galen's Prophecy: Temperament in Human Nature*, Basic Books, New York, 1994.

21 Hamer, DH, Hu, S, Magnuson, VL, Hu, N and Pattatucci, AML, 'A linkage between DNA markers on the X chromosome and male sexual orientation', *Science*, 261, 1993, 321–7.

22 Thanks to Jenny Kitzinger for passing these on to me.

23 Rice, G, Anderson, C, Risch, N and Ebers, G, 'Male homosexuality: absence of linkage to microsatellite markers at Xq28', *Science*, 284, 1999, 665–7.

24 Quoted in Breggin, PR, and Breggin, GR, *The War against Children of Color*, Common Courage Press, Monroe, ME, 1998.

25 Bock, GR and Goode, JA, eds, *Genetics of Criminal and Anti-Social Behaviour*, Ciba Foundation Symposium 194, Wiley, New York, 1996.

26 Brennan, PA, Mednick, SA and Jacobsen, B, 'Assessing the role of genetics in crime using adoption cohorts', in Bock, GR and Goode, JA, op. cit., pp. 115–22.

27 Brunner, HG, Nelen, M, Breakfield, XO, Ropers, HH and van Oost, BA, 'Abnormal behavior associated with a point mutation in the structural gene for monoamine oxidase A', *Science*, 262, 1993, 578–80.

28 Cases, O et al., 'Aggressive behavior and altered amounts of brain serotonin and norepinephrine in mice lacking MAOA', *Science*, 268, 1995, 1763–8.

29 Caspi, A and seven others, 'Role of genotype in the cycle of violence in maltreated children', *Science*, 297, 2002, 851–4.

30 Caspi, A and ten others, 'Influence of life stress on depression: moderation by a polymorphism in the 5-HTT gene', *Science*, 301, 2003, 386–9.

31 The science writer Matt Ridley has subsumed such interactions within the phrase *Nature through Nurture*, the title of his recent book, Fourth Estate, 2003.

32 Cavazzana-Calvo, M, Thrasher, A and Mavilio, F, 'The future of gene therapy', *Nature*, 427, 2004, 779–81.

33 Caplan, AL, 'Is better best?', *Scientific American*, 289, 2003, 104–5.

34 Chorney, MJ and eleven others, 'A quantitative trait locus associated with cognitive ability in children', *Psychological Science*, 9, 1998, 159–66.

35 Jamieson, JW, 'Concerning scientific creativity: Hermann J. Muller and germinal repositories', *Mankind Quarterly*, 33, 1993, 443.

36 In Rose, SPR, Lewontin, RC and Kamin, LJ, *Not in Our Genes*, Penguin, Harmondsworth, 1984.

37 Rose, SPR and Rose, HA, 'Do not adjust your mind; there is a fault in reality', *Cognition*, 2, 1974, 479–502.

38 Andreasen, N, *Brave New Brain: Conquering Mental Illness in the Era of the Genome*, Oxford University Press, New York, 2001.

39 Barondes, SH, *Better than Prozac: Creating the Next Generation of Psychiatric Drugs*, Oxford University Press, New York, 2003

40 Editorial, 'Better reading through brain research', *Nature Neuroscience*, 7, 2004, 1.

41 McGaugh, JL, *Memory and Emotion*, Weidenfeld and Nicolson, 2003.

42 Murphy, S, Hay, A and Rose, S, *No Fire, No Thunder*, Pluto Press, 1984.

43 Watson, P, *War on the Mind: The Military Uses and Abuses of Psychology*, Hutchinson, 1978.

44 George, MS, 'Stimulating the brain', *Scientific American*, September 2003, pp. 67–73.

45 McMurtrey, op. cit. The following four references are derived from that report.

46 'Surveillance Technology, 1976: policy and implications, an analysis and compendium of materials: a staff report of the Subcommittee on Constitutional Rights of the Committee of the Judiciary', United States Senate, Ninety-fourth Congress, second session, p. 1280, US GOV DOC Y 4.J 882:SU 7/6/976.

47 Oskar, KJ, 'Effects of low power microwaves on the local cerebral blood flow of conscious rats', Army Mobility Equipment Command Report, # AD-A090426, 1980. Available from NASA Technical Reports. Abstract at http://www.raven1.net/v2snasa.htm

48 Castelli, CJ, 'Questions linger about health effects of DoD's "Non-Lethal Ray",' *Inside the Navy*, 14 (12), 2001, 1–6 http://globalsecurity.org/org/news/2001/e20010327questions.htm and http://www.pjproject.org/usaf.html

49 Hodges, A, *Alan Turing: The Enigma of Intelligence*, Counterpoint, 1983.

50 Rose, S, *The Making of Memory*, Bantam, 1992, 2nd ed. Vintage, 2003.

51 Grand, S, *Growing up with Lucy: How to Build an Android in Twenty Easy Steps*, Weidenfeld and Nicolson, 2004.

52 Warwick, K, *I, Cyborg*, Century, 2002.

53 Baudrillard, J, *Les mots de passe*, Pauvert, Paris, 2000.

54 Stock, G, *Redesigning Humans: Choosing Our Children's Genes*, Profile Books, 2002.

55 Silver, L, *Remaking Eden: Cloning and Beyond in a Brave New World*, Weidenfeld and Nicolson, 1998.

56 Fukuyama, F, *Our Posthuman Future: Consequences of the Biotechnology Revolution*, Profile Books, 2002.

Chapter 12

1 Dana Foundation, *Neuroethics: Mapping the Field*, Dana Press, Washington DC, 2002.

2 Busquin, P et al., *Modern Biology and Visions of Humanity*, Multi-Science Publishing, Brussels, 2004.

3 Horgan, J, *The Undiscovered Mind: How the Brain Defies Explanation*, Weidenfeld and Nicolson, 1999.

4 Blank, RH, *Brain Policy: How the New Neuroscience Will Change Our Lives and Our Politics*, Georgetown University Press, Washington DC, 1999.

5 Rose, S, *Lifelines: Biology, Freedom, Determinism*, Penguin, 1997, 2nd ed. Vintage, forthcoming; Rees, D and Rose, S, eds, *The New Brain Sciences: Prospects and Perils* (especially the chapters by Stephen Sedley, Peter Lipton and Mary Midgley), Cambridge University Press, Cambridge, 2004.

6 Rosenberg, A, *Darwinism in Philosophy, Social Science and Policy*, Cambridge University Press, Cambridge, 2000.

7 Sedley, S, in Rees, D and Rose, S, eds, *The New Brain Sciences*, op. cit., pp. 123–80.

8 Reznek, L, *Evil or Ill? Justifying the Insanity Defence*, Routledge, 1997.

9 Crick, FHC, *The Astonishing Hypothesis: The Scientific Search for the Soul*, Simon and Schuster, 1994.

Index

Numbers in italics refer to Figures.